DISCOVERING INTELLIGENT DESIGN

WORKBOOK

Hallie Kemper
Casey Luskin
Gary Kemper

DISCOVERY INSTITUTE PRESS, SEATTLE, WA

Description

The *Discovering Intelligent Design Workbook* is part of a comprehensive curriculum that presents both the biological and cosmological evidence in support of the scientific theory of intelligent design. Developed for middle-school-age students to adults, the curriculum also includes a textbook and a DVD with video clips keyed to the content of the textbook. The workbook provides review questions, vocabulary questions, and essay questions to enhance the curriculum's educational value for students. The workbook also contains inquiry activities to give students hands-on opportunities to learn about intelligent design. These activities allow students to experimentally investigate questions like "Why does ice float?" or "What is the Doppler effect?," to critically analyze media coverage of the debate over intelligent design, and to even build their own "universe creating machine." Produced by Discovery Institute in conjunction with Illustra Media, the curriculum is intended for use by homeschools and private schools. More information can be obtained by visiting the curriculum's website, http://discoveringid.org.

Library Cataloging Data

Discovering Intelligent Design Workbook by Hallie Kemper, Casey Luskin, and Gary Kemper

148 pages, 8.5 x 11 inches

Library of Congress Control Number: 2013934289

BISAC: SCI015000 SCIENCE/Cosmology

BISAC: SCI027000 SCIENCE/Life Sciences/Evolution

BISAC: SCI034000 SCIENCE/History

BISAC: SCI075000 SCIENCE/Philosophy & Social Aspects

ISBN-13: 978-1-936599-09-7

ISBN-10: 1936599090

Publisher Information

Discovery Institute Press, 208 Columbia Street, Seattle, WA 98104

http://www.discoveryinstitutepress.com

Published in the United States of America on acid-free paper.

First Edition, First Printing, May 2013.

INTRODUCTION

WORKBOOK

This workbook is intended to accompany the *Discovering Intelligent Design* textbook. There is also a DVD which accompanies the textbook, but all of the questions and activities in this workbook pertain to the textbook. In fact, for the vast majority of the workbook questions, the answer can be found directly in the corresponding chapter in the textbook.

There are four different types of learning activities in the workbook:

- **Fill-in-the-blank or short answer questions,** which typically focus on defining or reviewing terms or concepts discussed in the textbook. Occasionally they may require some research.
- **Vocabulary questions,** which might be fill-in-the-blanks, matching definitions, or a crossword puzzle.
- **Essay questions,** which ask the student to write an answer exploring some topic in more detail.
- **Inquiry activities,** which allow the student to do some kind of hands-on and/or critical thinking activity to develop scientific thinking skills and improve learning.

The textbook and workbook are each divided into six sections:

- **Chapters 1–2 (Section I)** provide an overview of the viewpoints and terms common to the subject of intelligent design.
- **Chapters 3–6 (Section II)** examine theories about the origin and development of the cosmos.
- **Chapters 7–12 (Section III)** investigate the origin of life, including the origin of information in DNA, and also critique Darwinian evolution while presenting the case for biological design.
- **Chapters 13–17 (Section IV)** explore the evidence for common descent, and many problems with the theory that all living organisms are related.
- **Chapter 18 (Section V)** summarizes the evidence for design.
- **Chapters 19–20 (Section VI)** review the tactics used by critics to silence debate, respond to some common objections, and give ideas about how the reader can get further involved.

An answer section has been provided at the end of the workbook.

CHAPTER 1

DESIGN DECODED

CHAPTER 1: DEFINING MOMENT

1. _____ is the scientific theory that studies indications of design in nature.

2. Should the evidence for design in nature be controversial? Explain your answer:

_____

_____

_____

3. Materialism is the philosophical belief that

_____

_____

4. Define methodological naturalism:

_____

_____

5. Name two things that blind and unguided natural processes cannot do:

_____

_____

6. How do ID theorists detect indications of design in nature?

_____

_____

7. How is intelligent design different from creationism?

_____

_____

_____

_____

8. In the context of science and information theory, what does it mean to say something is complex?

_____

9. What does it mean to say something is specified?

_____

10. When complex and specified information is observed in nature, what can be inferred as its cause?

_____

11. For each of the items below, indicate yes or no.

Item	Complex	Specified
Salt Crystal		
Your hair when you wake up in the morning		
Morse code		
Ripples on Seashore		
Arrangement of garbage at the dump		
The pattern of ink on today's newspaper		

Fill in the blanks below with words from the box.

creationism	materialism
CSI	philosophy
intelligent agent	Socrates
intelligent design	tenet

12. _____ A being with the ability to plan ahead and think with an end-goal in mind.

13. _____ A principle held in common by members of a group.

14. _____ Some features of the universe and life are best explained by an intelligent cause.

15. _____ The belief that the material world is all that exists.

16. _____ Philosopher who urged us to "follow the argument wherever it leads."

17. _____ The belief that the universe and life were created by the God of the Bible.

18. _____ A set of beliefs about the nature of the world, and how one ought to live life.

19. _____ Complex and specified information, a hallmark of intelligent design.

ESSAY

Choose two or three tenets of materialism. Based on what you currently know, write your opinion on whether they are valid or invalid and why.

INQUIRY ACTIVITY: *Seeking Specified and Complex Information around the House*

Scientists explain that we can detect design when we find something that is both complex (i.e., unlikely) and specified (i.e., matches an independent pattern). Being just complex (e.g., a random splattering of paint on the wall), or just specified (e.g., the shape of a salt crystal) is not enough to infer design. This inquiry activity will help you identify everyday items that are complex, specified, or both.

- **Materials Needed:** Household items and items from the outdoors.

- **Step 1:** Take five items from your house or classroom, and five items from the outdoors. For each item, explain whether it is complex, specified, or both. Should you infer that the object was designed?

- **Step 2:** Not everything inside or outdoors may be best explained by intelligent design. For example, can you find something in or near your house that is specified but not complex? Likewise, is there something that is complex, but not specified? Explain whether intelligent design is the best explanation for what you found.

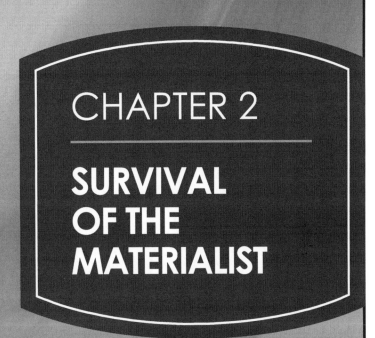

CHAPTER 2

SURVIVAL OF THE MATERIALIST

CHAPTER 2: SURVIVAL OF THE MATERIALIST

1. What two scientists are credited with the theory of evolution by natural selection?

   Charles Darwin and Alfred Russel Wallace.

2. What is the name of the famous book that first explained this theory?

   The Origin of the Species by Means of Natural Selection by Charles Darwin.

   When was it published? ___1859___

3. Natural selection was first proposed by comparing it to the process of

   The theory of the inheritance of aquired characteristics.

4. What 20th century discovery revealed problems with Lamarck's theory of inheritance of acquired characteristics?

   The discovery of genes.

5. Which type of evolution looks at "small-scale changes in a population of organisms"?

   Microevolution.

6. Which type of evolution claims that "all life forms descended from a single common ancestor through unguided natural processes," theorizing large-scale changes in populations of organisms?

   Macroevolution

7. Does microevolution demonstrate macroevolution? Explain your answer:

   No, the two are different. Microevolution demonstrates small changes over time, and the changes don't make the animal turn into a different species, on the contrary to Macroevolution.

8. What is taxonomic classification?

It is the grouping of organisms into categories based upon features considered similar by scientists.

9. What is the lowest level of taxonomic classification?

Species.

What is the highest level? Domain.

10. What is circular reasoning?

It is an invalid type of reasoning where the rules/ starting assumptions grant only the desired results.

11. What four critical thinking steps should you take as you investigate Darwinism and intelligent design?

- Question Assumptions
- Demand Evidence
- Define Your Terms
- Seek the Best Explanation

12. How many years has it been since Darwin's theory was first proposed? 1838
Is it appropriate to reevaluate evolutionary thinking? Explain why or why not:

I think it is. We should always look for new discoveries and allow them to challenge evolutionary thinking so we can get closer to the truth.

I would like to talk about that question

13. In what ways is intelligent design compatible with Darwinian evolution?

_____

_____

This too

14. How is ID incompatible with Darwinian evolution?

_____

_____

Fill in the blanks below with words from the box.

artificial selection	macroevolution
circular reasoning	microevolution
Darwinism (original theory)	natural selection
evolutionist	neo-Darwinism
Lamarck	seek more evidence

15. _Seek more evidence_ If there are two or more possible explanations, you should do this.

16. _Artificial Selection_ The selection of certain animals by a breeder in order to increase or eliminate traits in a population.

17. _Darwinism (original)_ All life evolved from a single common ancestor through descent with modification, driven by unguided natural selection acting upon random variation.

18. _Microevolution_ Small-scale changes within a species.

19. _Neo-Darwinism_ All life evolved from a single common ancestor through descent with modification, driven by unguided natural selection acting upon random DNA mutations.

20. _Circular Reasoning_ A form of reasoning where the rules or starting assumptions permit only the desired results.

21. _Natural Selection_ An unguided natural process where organisms better suited to survive and reproduce will tend to pass on their traits to offspring.

22. _Macroevolution_ Large-scale biological changes leading to fundamentally new biological features.

23. _Lamarck_ Developed theory of inheritance of acquired characteristics.

24. _Evolutionist_ A person who believes in macroevolution.

ESSAY - _Don't do this_

In Chapter 2, we described thinking tools such as questioning assumptions, demanding answers, defining your terms, and seeking the best explanation. What are some of the benefits of using these tools to evaluate the debate over intelligent design and materialism?

INQUIRY ACTIVITY: _Develop Critical Thinking Skills_

Critical thinking is necessary for understanding the complex world around us. One of the best ways to gain critical thinking skills is to study the arguments and viewpoints that other people are advocating—and then analyze them. You can do this by looking at the opinion section of your local newspaper.

- **Materials Needed:** Newspaper.

- **Step 1:** Find the opinion section. (Note: Sunday editions of newspapers often print more opinion articles and editorials than on most other days of the week.)

- **Step 2:** Find two editorials or opinion articles in the paper, one you agree with and one you disagree with. For each article, apply the four critical thinking tips offered in Chapter 2:

(a) Question Assumptions: Identify assumptions in the article's argument. Explain whether those assumptions are valid.

(b) Demand Evidence: Identify instances where the article provides evidence to back up its claims. Are there any examples where the article makes assertions that are not backed up by evidence?

(c) Define Terms: Does the article use any complex terms that might have different meanings, or might be unfamiliar to some readers? Does the editorial define those terms? How might this confuse or mislead the reader?

(d) Seek the Best Explanation: Evaluate the argument made in the editorial. Does the author make a persuasive case? Why or why not?

CHAPTER 3

THINK BIG...
NO, BIGGER

9/19/16

1. How many miles does light travel in a year?

   5.88 trillion miles.

2. Complete the *kalam* argument:

   1. Anything that begins to exist has a cause

   2. The universe began to exist

   3. Therefore... _The universe has a cause_

3. Describe two problems that the Big Bang theory poses for materialism:

   - _It shows that the universe had a begining._

   - _Cosmic fine-tuning_

4. Name the theory developed by Albert Einstein that is a description of gravity as it relates to space and time: _General Relativity_

5. What was Einstein's error regarding the cosmological constant?

   He became convinced the universe was not eternal and the argument didn't allow for that.

6. Cosmologists theorized that _if the universe began with a Big Bang there_ would be left after the Big Bang. Precise measurements from the _COBE satellite_ reinforced the theory in the early 1990s.

7. At least _100 billion ?_ galaxies are estimated to exist in our observable universe.

8. What is the redshift, and how did Edwin Hubble use it to confirm that galaxies are receding from one other?

Redshift is when light waves coming from a receding object are stretched to a lower frequency, and thus shifted down towards the red end of the color spectrum. Hubble measured the redshift of the galaxies and confirmed they were receding.

9. Describe the properties of the three main cosmological models:

	infinitely old or finite age?	expanding or constant size?
Static Model	eternal	constant
Steady State Model	finite	expanding
Big Bang Model	old	expanding

10. Which cosmological model is best supported by the scientific evidence:

A. Static   B. Stead State model   C. Big Bang Model

11. Can you think of anything that began to exist that did not have a cause?

No.

Match the words with their meaning:

12. _g._ First person to propose that the universe began with the explosion of a primeval atom.

13. _a._ The study of celestial bodies and phenomena.

14. _d._ Explains the frequency of sound waves emitted by moving objects.

15. _e._ A description of gravity as it relates to space and time.

16. _h._ A logical argument that concludes the universe had a first cause.

17. _c._ The study of the universe as a whole.

18. _b._ The name given to the event that began the universe.

19. _j._ The cosmological theory that attempted to explain the expansion of the universe while still supporting the eternal universe hypothesis.

20. _i._ Ancient philosopher who argued that the universe was designed.

21. _f._ His research showing galaxies are receding from one another helped expose Einstein's greatest blunder.

> a. astronomy
> b. Big Bang
> c. cosmology
> d. Doppler effect
> e. General Relativity
> f. Edwin Hubble
> g. Georges Lemaître
> h. *kalam* argument
> i. Plato
> j. Steady State

ESSAY

Describe your views about the validity of the Big Bang cosmological model, including its impact on the philosophy of materialism.

INQUIRY ACTIVITY: *The Doppler Effect*

In Chapter 3 we learned about the evidence supporting the Big Bang model of the origin of the universe. One of those lines of evidence is the redshift in the frequency of light coming from other galaxies. Because of the Doppler effect, light coming from a receding object will have a lower frequency, and light coming from an object moving closer will appear to have a higher frequency.

Though caused by different physical principles, a similar effect is observed with sound. Sound coming from a receding object will have a lower pitch (frequency), whereas the pitch of sound coming from an object moving closer will be higher. This inquiry activity will explore the Doppler effect with sound.

- **Materials Needed:** Three people; a Nerf ball or large foam-ball; a kitchen knife; duct tape; a small battery-operated electric buzzer; a battery compartment; batteries; and a large outdoor space.

  > Note: Stores like Radio Shack are good places to find a battery-operated electric buzzer and a battery compartment. Stores like Target are good places to buy a Nerf Ball or other large foam-ball.

- **Step 1:** With the kitchen knife, *carefully* cut a slit in the Nerf ball (or other foam ball) large enough to insert the electric buzzer inside.

- **Step 2:** Connect the wires of the battery compartment to the wires on the electric buzzer. (Red wires connect to red wires, and black wires connect to black wires.) Insert the batteries into the battery compartment. The buzzer should now be buzzing.

- **Step 3:** Insert the buzzing buzzer into the slit in the foam ball. Tape the hole shut with duct-tape, so the buzzer will not come out when the ball is thrown. You've now made a "buzzer-ball."

- **Step 4:** Position two people (the throwers) in the large outdoor area, about 15 to 20 feet away from one another. Position the third person (the observer) between the other two, but off to the side by about 5 to 10 feet.

- **Step 5:** The throwers should now throw the ball back-and-forth to one another. The observer should observe the pitch of the buzzer-ball as it moves toward, and away, from them. Did the pitch of the sound coming from the buzzer-ball change as the ball goes by? How does it change?

CHAPTER 4

DON'T TOUCH THAT DIAL

10/3/16

1. The universal constants and laws of physics indicate design because they are

to finely tuned _____ for complex life.

2. What is Dark Energy?

The effect in the universe that acts to push galaxies away from one another, accelerating the expansion of the universe.

3. Name the force that scientists believe acts to slow down the expansion of the universe:

Gravity.

4. Name three of the many finely tuned factors needed for the universe to sustain complex life:

- *Strong nuclear force*
- *Electromagnetic force*
- *Gravitational Constant*

5. Scientists believe that many elements necessary for life (e.g., iron, carbon, or oxygen) were manufactured within

Stars.

6. The fine-tuning of the cosmological constant has been compared to the probability of a dart being thrown at a dartboard the size of our galaxy, and hitting a bull's-eye less than the size of a:

Quarter.

7. What happens to the elements in a star when it dies?

They scatter across the ~~universe~~ galaxy.

8. Approximately how many generations of stars had to be born and die in order to generate the elements necessary for life?

Three generations, adding up to at least 9 billion years.

9. How old do most scientists believe the universe is?

13.7 billion years old.

10. The initial arrangement of mass and energy in the universe is known as the

Big Bang. Initial Entropy

11. Place the following items in order of greatest to least magnitude of size:

~~Solar system~~	~~Universe~~	~~Moon~~	~~Jupiter~~
~~Earth~~	~~Galaxy~~	~~Sun~~	

a. _Universe_

b. _Galaxy_

c. _Solar System_

d. _Sun_

e. _Jupiter_

f. _Earth_

g. _Moon_

Fill in the blanks with words from the box.

12. _Dark Energy_ — The effect that pushes celestial objects away from each other, accelerating the expansion of the universe.

13. _Strong Nuclear Force_ — A force which, if weaker, would cause hydrogen to be the only element.

14. _Gravitational Constant_ — If this factor were even slightly smaller, stars would never burn and produce heavy elements.

15. _Cosmological Constant_ — This is a measure of dark energy's effect on the expansion of the universe.

16. _Gravity_ — This force causes matter to attract.

17. _Sufficient_ — Means, "enough without needing anything else."

18. _Electromagnetic Force_ — If the strength of this force were changed, atomic bonds could not form.

19. _Necessary_ — Means, "required, but maybe not enough by itself."

20. _Initial Entrophy_ — A parameter that is extremely fine-tuned to enable matter in the universe to condense into stars and galaxies.

ESSAY *do next week (week 8)*

Given the fine-tuning of the universe, does the evidence support materialism's tenet #2 ("The physical laws and constants of the universe ultimately occurred by purposeless, chance processes") or does it indicate intelligent design? Explain your answer.

INQUIRY ACTIVITY: *Build a Universe-Creating Machine*

For this inquiry activity, you're going to build a universe-creating machine (well, not a real one) that shows the values of various physical laws and constants necessary for a life-friendly universe. You can use your artistic creativity to design the machine however you like.

- **Materials Needed:** List of finely tuned parameters; cardboard shoebox (suggested); pushpins; and lots of creativity.

- **Step 1:** Review Chapter 4 and/or search online for a list of physical parameters that must be finely tuned for life to exist. Choose a few parameters to include in your universe-creating machine.

- **Step 2:** Now that you've identified a few finely tuned parameters, try to find the precise values those parameters must have to allow for life. We'll give you one example here: Chapter 4 lists the Gravitational Constant as a finely tuned parameter, though we haven't given the exact value. What you need to do is go online and search for the exact value of the Gravitational Constant. A quick search of "Gravitational Constant" shows its value is: 6.673×10^{-11} m^3 / kg / s^2.

  Note: Some parameters might not have a specific value. For example, Chapter 4 listed fundamental forces, such as the weak and strong nuclear forces, and the electromagnetic force. Since those are "forces," they will have different strengths depending on the physical circumstances. One way to get a precise number measuring the strength of the force is to find the value of the "coupling constant" involved in those forces. Thus, searching for the "fundamental force coupling constants" will give you values you can use in your universe-creating machine. We've now given you hints on how to find the values for at least four parameters.

- **Step 3:** Build a universe-creating machine! You can build a machine that looks something like the "Universe-Creating Machine" illustration in Chapter 4 (Figure 4-3), or you can use your own creativity to build one that's completely different. The only requirements are as follows:

- The machine must have a dial for each finely tuned parameter you researched.
- Each dial should be able to turn, and must have numbers on it representing different possible values of the parameter.
- You should make a mark on the dial indicating the value that is "life-friendly." An example of what the dial might look like for the Gravitational Constant is seen below:

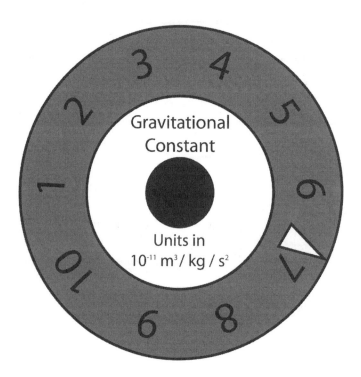

- Using pushpins, attach the dials to the outside of the shoebox.

You now have a mock universe-creating machine. Careful—don't change the settings needed for life!

CHAPTER 5

THE EMPIRE STRIKES BACK

10/23/16

In questions 1-5, name the alternative explanations that materialists use to avoid the implications of the Big Bang:

1. There are an infinite number of universes, each with different laws and constants, and we occupy the one that got lucky and allows complex life:

   Multiverse Theory

2. A fluctuation in the void spawned the universe:

   Chance Universe

3. The universe would have to exist before it was created:

   Self - creation

4. The universe popped into existence in the same way that subatomic waves and particles do:

   Quantum Theory

5. The theory that the universe perpetually expands, then collapses, and then expands again:

   Oscillating Universe Theory

Describe the rebuttals to the claims in questions 6-10:

6. Multiple-Universe (Multiverse) Theory:

   • There is no explanation for the cause of the multiverse.
   • The theory relies on the assumption that the universes would be different from each other. If there are multiple universes, why wouldn't many, or all of them be the same?
   • Since we cannot observe anything outside our universe, these theories are 100% philosophical speculation, not science.

7. Oscillating Universe Theory:

• There's no known physical mechanism for making a cyclic "bounce" making it impossible for a univers to pass through a singularity.

8. Comparing the origin of the universe to the origin of quanta:

Quantum events occure within a range of well-defined statistical probabilities according to the quantum theory. This suggests that quanta has an underlying cause.

9. Chance Universe:

• What was it that fluctuated?

• What caused that non-existant something to fluctuate?

• Why was there an environment that allowed for such "fluctuations"?

10. Self-Creating Universe:

• For anything to create itself, it would have to exist before it was created. Most people would agree that this is logically absurd.

11. Vocabulary Puzzle

This is not a normal word puzzle. In addition to occurring forward, backward, horizontally, vertically, or diagonally, words in this puzzle might wrap around one or more angles. The letters just have to touch. For example, the word "follow" can be found as seen in the puzzle at the right:

```
S  F  B  A  I
J  O  I  W  M
E  L  L  T  S
Q  B  O  O  V
V  W  N  P  A
```

A few additional rules: Some letters may be used for multiple words, but no letter will be used more than once in the same word. Also, the two boxes with letters around them are not used in any word.

Finally, after all the words have been found, read the remaining letters in normal left-to-right order and you'll discover a motto that should be used by anyone reading this book.

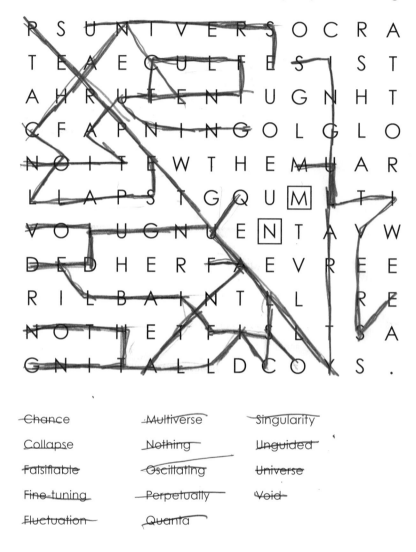

Chance	Multiverse	Singularity
Collapse	Nothing	Unguided
Falsifiable	Oscillating	Universe
Fine-tuning	Perpetually	Void
Fluctuation	Quanta	

The phrase: *I can't read it*

ESSAY

Which, if any, of the materialist scenarios claiming an unguided origin of the universe do you find to be the most plausible? Explain the arguments for and against it.

INQUIRY ACTIVITY: *Ruling Out Chance*

In Chapter 5, we learned that the multiverse hypothesis allows materialists to invent untold numbers of unobserved universes in order to explain highly unlikely events, like the fine-tuning of the universe. In real science, however, scientists must be able to test between random/chance events and regular patterns.

Normally, scientists will form a hypothesis and then make predictions about what type of pattern ought to appear if their hypothesis is correct. If the data appear to be random, then they reject their hypothesis. But if a non-random pattern appears, they can rule out chance and try to determine the best explanation for the observed data.

Multiverse thinking makes it effectively impossible for scientists to rule out chance, because materialists can just invent more universes until the pattern they observe would happen by chance. Such thinking destroys our ability to do science.

The purpose of this inquiry activity is to help you learn to discriminate between random data and real patterns. The activity requires two people: one to be the "Dealer" and one to be the "Investigator." The Dealer gets to decide how the deck of cards is going to be ordered, and how many cards will be dealt. The Dealer will then deal those cards out to the Investigator. The Investigator's job is to determine if there's some pattern to the cards that can't be explained by chance, or if the Dealer simply shuffled the deck at random.

- **Materials Needed:** A normal deck of 52 cards and two people.

- **Step 1:** The Investigator leaves the room.

- **Step 2:** The Dealer arranges the deck of cards in one of two ways, either (a) shuffling the cards so they are arranged randomly, or (b) carefully ordering the deck of cards so the cards appear according to a particular pattern. There are many possible patterns the Dealer could use; here are a couple of ideas:

- Red suit, Black suit, Red suit, Black suit, etc.
- 2, 3, 4, 5, 6, 7, 8, 9, 10, Jack, Queen, King, Ace, etc.
- Heart, Diamond, Spade, Club, Heart, Diamond, Spade, Club, etc.

- **Step 3:** The Investigator comes back into the room and the Dealer deals the cards.

- **Step 4:** The Investigator studies the cards and tries to determine if they're randomly shuffled, or if there is a pattern.

- **Step 5:** The Investigator tells the Dealer his/her hypothesis about how the Dealer arranged cards. The Dealer then tells the Investigator if he/she was correct. Feel free to switch roles and play multiple times.

CHAPTER 6

HOME, SWEET HOME

10/23/16

OR

1. Name the belief that we occupy no special place in our galaxy:

Copernican Principle

2. Name four of the many parameters about our planet that make it specially suited for life:

X _Its Location in the Galaxy liquid iron core_

X _Its location in the solar system liquid water_

• _A large moon_

• _Plate tectonics_

also - type of star; stable orbit; terrestrial planet
galactic habitable zone

X 3. The _liquid iron core_ is the location that avoids deadly
radiation at the galactic core, yet contains the elements necessary for complex life.

4. The _circumstellar habitable zone_ is the location in our solar system
that permits temperatures that are just right for liquid water.

5. These two giant planets in our solar system help to protect Earth from asteroids:

Jupiter and _Saturn_

6. Jupiter's diameter is _more than 11_ times larger than that of our planet.

7. What event is theorized to have caused the formation of our moon?

Supposedly a planet roughly the size of Mars collided w/
earth and some of the debris collected together to form the moon

8. Explain the importance of our moon to life on Earth:

Its gravitational pull stabilizes the tilting of Earth's axis at 23.5 degrees, guaranteeing relatively mild seasonal changes.

9. Name the 16th century mathematician who theorized that Earth was not the center of the universe?

Nicolaus Copernicus

10. What two main gasses comprise Earth's atmosphere? *Nitrogen*

and *Oxygen*

11. Give two reasons why liquid water is important for life on Earth:

- *Water is essential for life as a sustence to drink and gain nutrients from.*
- *Most of Earth's life resides in the ocean, so it is also necessary as a habitat. Better answer is: It has a high surface tension, fostering many biological processes.*

Match the words with their meaning:

12. *b.* The belief that our planet is relatively insignificant.

13. *f.* Research program looking for advanced life beyond Earth.

14. *e.* The galaxy containing our solar system.

15. *h.* Land-based.

16. *a.* Surrounding a star.

17. *d.* The area in a galaxy that is favorable to life.

18. *g.* Forces on Earth that cause movement within its crust.

19. *c.* This won't happen because of water's unique property of being denser as a liquid than as a solid.

a. circumstellar
b. Copernican Principle
c. frozen oceans
d. galactic habitable zone
e. Milky Way
f. SETI
g. tectonic
h. terrestrial

20. Fill in the blanks for the solar system diagram below.

ESSAY

Do you think the Copernican Principle is valid, or is the Earth a privileged planet? Explain.

INQUIRY ACTIVITY: *Why Does Ice Float?*

One finely tuned parameter mentioned in Chapter 6 is that water is less dense in its solid form than as a liquid. If water did not have this unique property, ice would sink, freezing the oceans from the bottom up. Life as we know it could not exist. This inquiry activity will help you explore the density of solid versus liquid water, and understand why ice floats. Remember, density = mass divided by volume, and can be measured in grams / milliliter, or g / cm^3.

Before we begin, we should give a preliminary note: This inquiry lab requires some very basic chemistry equipment, and you won't be able to do the lab easily without that equipment. Specifically, the lab requires a graduated cylinder and a scale for use in simple chemistry experiments. If you don't have access to those, you should be able to find them at any teacher's supply store. Without them, this activity will be difficult.

- **Materials Needed:** A graduated cylinder with marks for measuring volume; ice cubes; liquid water; a high-quality scale capable of measuring to the nearest tenth of a gram (e.g., a triple-beam balance or electronic scale for chemistry labs); tongs; a pencil or paper clip.

For each step, record your observations in the table below:

1: Initial Mass of Graduated Cylinder When Empty	
2: Volume of Water (without ice)	
3: Mass of Liquid Water and Graduated Cylinder	
4: Mass of Liquid Water	
5: Density of Liquid Water	
6: Volume of Liquid Water in Graduated Cylinder	
7: Mass of Ice Cube Alone on the Scale	
8: Volume of Water in Graduated Cylinder with Ice Cube	
9: Volume of Ice Cube.	
10: Density of Ice Cube.	

- **Step 1:** Find the mass of the graduated cylinder when empty. Record mass in Row 1.

- **Step 2:** Partially fill the graduated cylinder with liquid water. Record the volume of the water in Row 2.

- **Step 3:** Find the mass of the graduated cylinder with liquid water. Record mass in Row 3.

- **Step 4:** Calculate the mass of liquid water (Row 3 minus Row 1). Record in Row 4.

- **Step 5:** Calculate the density of liquid water (Row 4 divided by Row 2). Record in Row 5. Include units in your final answer.

- **Step 6:** Empty the graduated cylinder, and refill it with water. Record the volume of the water in Row 6.

- **Step 7:** Take an ice cube (or part of an ice cube) that can fit in the graduated cylinder. Be sure to only touch it using tongs so it doesn't melt.

- **Step 8:** Find the mass of the ice cube alone on the scale. Record in Row 7.

- **Step 9:** Pick up the ice cube with tongs and place it in the graduated cylinder.

- **Step 10:** Take the pencil or paper clip and push the ice cube so it sits just below the surface of the water.

- **Step 11:** Record the volume of the water with the ice cube in it. Record in Row 8.

- **Step 12:** Calculate the volume of the ice cube (Row 8 minus Row 6). Record in Row 9.

- **Step 13:** Calculate the density of the ice cube (Row 7 divided by Row 9). Record in Row 10.

- **Step 14:** Compare the density of liquid water (Row 5) to the density of ice (Row 10).

Why is the density of ice lower than that of liquid water? It has to do with the physical structure of ice crystals, and how water molecules pack together in a solid state. In a liquid state, water molecules pack more closely than they do in a solid state.

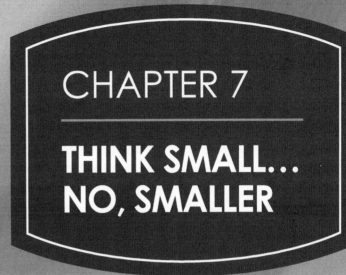

CHAPTER 7

THINK SMALL...
NO, SMALLER

CHAPTER 7: THINK SMALL... NO, SMALLER

1. What is spontaneous generation? _The sudden development of organisms from non-living matter. In essence, life can spontaneously come from non-life._

2. _Proteins_ are the worker molecules in a living cell and are made up of long chains of amino acids.

3. Define chemical evolution: _The theory that chemicals in nature assembled through blind, unguided, chance chemical reactions to create life._

4. How many different amino acids are generally used in living organisms to make proteins?
   _20_

5. Name three of the systems required in all living cells (e.g., protein production):
   - _Replication_
   - _Disposal of waste_
   - _Transportation of parts within the cell_

6. Which argument against the Miller-Urey experiment do you find to be the strongest and why?
   The oxidation argument. Because it says that there was enough free oxygen in the Earth's early atmosphere to create oxidation, which would destroy the organic compounds in the "prebiotic soup".

7. In the cell/book analogy, proteins are equivalent to what part of a book?
   Words & Sentences

8. ATP can be generated by a molecular machine called

ATP synthase

9. Who disproved the theory of spontaneous generation?

Louis Pasteur

10. Haldane's "prebiotic soup" is similar to Darwin's phrase:

"warm little pond"

11. Did free oxygen exist in early Earth's atmosphere? _Yes_

   Why is that important?

   There was enough to have caused immediate oxidation,
   which would destroy any organic compounds in the
   "prebiotic soup".

12. Give two reasons why the plasma membrane is important:

   • _Without it cells would be prone to harmful chemical reactions_
     and molecules.
   • _Keeps the cells components together to allow for cellular processes_
     to take place.

Fill in the blanks with words from the box.

amino acid	plasma membrane
ATP	primordial soup
chemical evolution	proteins
Miller-Urey experiment	ribosome
oxidation	spontaneous generation

13. _Spontaneous generation_ — Theory that life could arise suddenly from inanimate objects.

14. _Primordial soup_ — A hypothetical sea of simple molecules from which life formed by chance.

15. _Oxidation_ — The chemical combination of a substance with oxygen.

16. _Miller-Urey experiment_ — An experiment intended to show the possibility of chemical evolution.

17. _Amino acid_ — An organic compound that is the basic building block of a protein.

18. _Proteins_ — The worker molecules of the cell.

19. _Plasma membrane_ — Protective barrier found in all living cells.

20. _Chemical evolution_ — Theory that unguided chemical reactions in nature created life.

21. _Ribosome_ — Multi-part machine responsible for translating genetic instructions to make proteins.

22. _ATP_ — An energy-carrying molecule in all cells.

ESSAY

The third tenet of materialism holds that "life originated from inorganic material through blind, chance-based processes." After learning about the complexity of a single living cell and the systems within it, does this tenet seem realistic? Explain your answer.

INQUIRY ACTIVITY: *Spontaneous Generation*

In Chapter 7, we learned about the now-rejected idea of spontaneous generation. This activity will allow you to test that concept to see if maggots spontaneously form on meat that is sealed off. We will roughly follow an experiment that tested spontaneous generation conducted by the Italian naturalist Francesco Redi in the 1600s. Unfortunately, Redi was ahead of his time and few believed his results. The disproof of spontaneous generation was not widely accepted until Louis Pasteur's experiments in the 1800s.

- **Materials Needed:** Three large clear glass beakers or jars (each of the same size); ground beef (raw); plastic wrap; and paper towels (or some other porous material).

- **Step 1:** Put an equal amount of ground beef at the bottom of each of the glass beakers.

- **Step 2:** Cover one beaker with the paper towel, secured with a rubber band around the top.

- **Step 3:** Cover the second beaker with plastic wrap, also secured with a rubber band.

- **Step 4:** Leave the third beaker open.

- **Step 5:** Let the beakers sit for one to two weeks. Be forewarned, this experiment can get stinky. The beakers may be left outdoors or indoors, but whatever their environment is, they must:

  — be exposed to the possibility of flies
  — sit in a non-freezing temperature
  — not be exposed directly to rain

- **Step 6:** Each day, inspect the beef for maggots (without disturbing the coverings). Record your observations.

- **Step 7:** At the end of two weeks, make final observations about the three beakers. Were there maggots in any beakers? Were there differences between the beakers? What do you think accounts for those differences?

CHAPTER 8

INFORMATION, PLEASE

ok

1. What is a gene? _A gene is a basic unit of heredity, typically understood as a section of DNA that contains assembly instructions for a particular protein._

2. In the "chicken or the egg" question regarding the origin of life, DNA is one of three requirements. What are the other two?

* _enzymes_
* _cell membrane_

3. The letters A, C, T, and G signify _nucleotide_ bases in the DNA.

4. What do you think is the strongest argument against the RNA world hypothesis?

RNA can't fulfill the roles of proteins.

5. The sequential ordering of base pairs in DNA instructs cellular machinery to

link amino acids in the correct order to produce functional proteins.

6. Within the cell, transcription is the first step in the process of _x gene expression. making proteins_

7. What is the second step in the process?

Translation.

8. Elaborate on what occurs during the two basic steps involved in the building of a protein:

The first step is transcription - DNA instructions are copied into mRNA. The mRNA molecule travels to ribosome for translation which constructs the protein thru a link chain of amino acids.

9. In your opinion, could the information in living cells appear by chance? Why or why not?

No. The information is too complicated to appear suddenly by chance, so much so that scientists can't recreate a working cell. If we couldn't do it deliberately how could it happen by chance?

Fill in the blanks with words from the box.

A, G, C, T	RNA world
DNA	Transcription
Genetic Code	Translation
Genome	tRNA
mRNA	

10. ___DNA genome___ The full complement of genetic information in an organism.

11. ___Genome tRNA___ Molecule that ferries amino acids to the ribosome.

12. ___a, A, C, I___ The "letters" in the genetic code.

13. ___Genetic Code___ Set of rules used by cells to convert the genetic information in DNA or RNA into proteins.

14. ___Transcription___ Name of the first step in the process of making a protein.

15. ___RNA World___ Hypothesis that the first life form might have been composed solely of RNA.

16. _~~mRNA~~ DNA_ — The primary information-storage molecule found in all living organisms.

17. _~~tRNA~~ mRNA_ — A molecule that transports the information in DNA to the ribosome for manufacturing a protein.

18. _Translation_ — The step in the protein-making process where the ribosome assembles the amino acid chain.

ESSAY

Information is all around us. No one debates that the information we see on television, the Internet, or in a book originated by intelligent design. Describe how information is used in living cells.

INQUIRY ACTIVITY: *The Origin of Functional Information*

Which mechanism is more efficient at generating functional information: random chance or intelligence? This activity will help you understand the origin of useful complex and specified information.

- **Materials Needed:** Set of Scrabble tiles; a large bowl with a lid; open space on the floor.

- **Step 1:** Pour the scrabble tiles into a bowl, put a lid on it, and shake it up.

- **Step 2:** Close your eyes and pour the bowl full of Scrabble tiles on the floor.

- **Step 3:** Study the tiles on the floor, but leave them where they fell. Have any English words appeared? Do they form complete sentences? Record any meaningful information that arose.

- **Step 4:** Repeat Steps 1, 2, and 3 five times.

- **Step 5:** Pour the tiles on the floor, but then intelligently arrange the Scrabble tiles into a sentence.

- **Step 6:** Which method was more efficient at generating functional information: random chance or arrangement by intelligence? Explain why you think you obtained this result.

CHAPTER 9

OPENING THE BLACK BOX

11/27/16

1. Complete Darwin's test of evolution: "If it could be demonstrated that any complex organ existed, which could not possibly have been formed by numerous, successive, slight modifications... my theory would absolutely break down.

2. If a mutation neither increases nor decreases the likelihood of survival or reproduction, what is the result?

   Natural selection has no effect; it has no reason to preserve the change.

3. If a mutation decreases the ability to survive and/or reproduce, what is the result?

   Mutation would tend not to be preserved.

4. What gives hemoglobin its oxygen-attracting properties?

   Iron. More appropriate - It's special shape allows it to carry iron atoms that attract oxygen.

5. Define irreducible complexity:

   A single system that is made of many interacting parts. If any one of the parts are removed the system stops working.

6. Describe the difficulty that irreducible complexity poses for Darwinism:

   Darwinism relies on mutation, but Irreducible complex structures cannot evolve in a step-by-step fashion because they don't function until all parts are present & functional.

7. Describe the "hand-in-glove" fit in protein-protein interactions, and explain why this is important:

Proteins must connect through a "hand-in-glove" fit in order to accomplish their cellular functions. Amino acids in each protein must be specifically ~~fit~~ arranged to give it the proper shape for such a fit.

8. The odds of a chain of 150 random amino acids yielding a stable, functional protein fold are 1 in

10^{74}

9. Explain a problem that evolution runs into when multiple mutations are needed to produce a benefit: Most mutations are detrimental to organisms and it's very unlikely a benefitial one will pop into existance let alone many. Better answer: In many multi-cellular organisms ~~acres~~ evolving a modestly complex multi-mutation feature would require greater population sizes & more time than would be available over Earth's history

10. What is the function of a bacterial flagellum?

The flagellum serves as a propeller to move the bacteria.

What would be the consequence to a bacterium if it lacked functional flagella?

They wouldn't be able to find food because they wouldn't be able to move, to food or a hospitable living environment.

11. Could the flagellum have evolved without planning or design? Elaborate on your answer.

No. Even a flagellum is very ~~precise~~ precise and complicated. Chance just can't create something like that.

Match the words with their meanings.

12. _b_ To take and use for another purpose.

13. _e_ Oxygen-carrying protein.

14. _d_ A matching connection between two or more unevenly shaped parts.

15. _f_ Requires many mutations before providing a beneficial function.

16. _c_ Rotary engine providing propulsion for many bacteria.

17. _g_ Something that scientists find interesting but don't fully understand.

18. _a_ A system of interacting parts wherein the removal of any part causes loss of function.

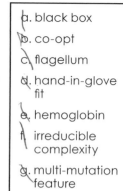

a. black box
b. co-opt
c. flagellum
d. hand-in-glove fit
e. hemoglobin
f. irreducible complexity
g. multi-mutation feature

ESSAY

Describe the basic problem for Darwinism when evolving stable protein folds, a hand-in-glove fit between enzymes, multi-mutation features, and irreducibly complex systems. How does this affect the scientific debate over ID and Darwinian evolution?

INQUIRY ACTIVITY: *Testing Random Mutation and Natural Selection*

In Chapter 9, we learned why the Darwinian mutation-selection mechanism has difficulty producing new functional biological information. Many theorists have attempted to create computer simulations of random mutation and natural selection to test its information-generating abilities. In this activity we'll use an online simulation, "The Richard Dawkins Mutation Challenge," to explore whether Darwinian processes can generate new information.

The simulation will pit you against the well-known evolutionary biologist and materialist Richard Dawkins. In his 1986 book *The Blind Watchmaker*, Dawkins offered a computer simulation of mutation and selection where he evolved the Shakespearean phrase "METHINKS IT IS LIKE A WEASEL."

Dawkins claimed his simulation properly modeled Darwinian evolution, but it didn't:

- It was impossible for the simulation to experience a harmful (deleterious) or lethal mutation, and thus it would simply keep mutating until eventually it found the "right" information sequence.

- Once it found a "right" mutation, it could never mutate into a wrong one.

- It didn't attempt to operate according to the rules of the genetic code and biology—it mutated the 26 letters in our alphabet, not the 64 codons in DNA that specify the amino acids used in life.

Unlike Dawkins, you will simulate evolution in an environment that works something like biology, using a programmed version of the actual genetic code. Your goal is to race against Richard Dawkins to see who evolves their target first.

You only have to evolve two specific amino acids using the rules of real biology, but he has to evolve the entire phrase "METHINKS IT IS LIKE A WEASEL," using the rules of his contrived simulation. Who will reach their target first? Play "The Richard Dawkins Mutation Challenge" to find out—and you'll quickly see how hard it is for Darwinian evolution to evolve just two simultaneous mutations under biologically realistic conditions.

- **Materials Needed:** A computer with an Internet connection.

- **Step 1:** Log on to the Internet and go to **http://www.mutationworks.com**

- **Step 2:** Read the instructions on the main page.

- **Step 3:** Follow the instructions to select your two starting amino acids, and your two target amino acids.

- **Step 4:** Start playing the simulation by clicking "Begin Challenge." Note: Initially, we recommend using one generation per year so you can easily see how many generations (i.e., years) it takes to evolve a multi-mutation feature.

- **Step 5:** We recommend playing around on the simulation for a while, as it may take a few minutes to become familiar with it. Have fun and good luck against Dawkins.

Do you think the simulation proposed by Dawkins represents a fair evaluation of neo-Darwinism?

CHAPTER 10

LIFE IS COMPLICATED

12/4/16

1. Consider the enormous number of differences between a bacterium and dog. List nine:

Intelligence	Needs	Purpose
Instinct	Flagellum	Brain
Size	Fur	Genetic Makeup

2. Fill in the blanks in the chart at right to complete the progression of parts that make up an organism:

Body Plan

Organs

Tissues

Cells

Proteins

Amino Acids

3. What are the functions of the plant's cell wall?

• Limits cellular growth
• Restricts movement
• Provides more protection
• Structural stability   more scientific answer:

4. Describe the process of photosynthesis: Chloroplasts capture energy from the sun using chlorophyll. Plants and other organisms convert light energy into chemical energy that can be later released to fuel the organisms. Chloroplasts use this energy to combine water & CO_2 to create oxygen & sugars = plant food.

5. Define metamorphosis: _A process of pre-programmed development where an organism changes its body plan._

What is the biggest problem it presents for materialism?

It has no real explanation on how a catapillar dissolves itself in the cuccoon (cacoon) and reasembles into a butterfly. I like your more involved answer than the answer key :)

6. What is an adaptation? _It is a feature that enables an organism to survive and reproduce in its environment._

7. List the basic types of tissues:

ANIMAL	PLANT
Nervous	ground
Muscle	dermal
Ephithelial	vascular
Connective	

8. _Connective_ (Epithelial) tissue surrounds the internal organs and lines the internal cavities of an animal.

9. What tissue type is used to cover the exterior of a plant? _Dermal tissue_

10. What tissue type transmits electrical signals within an animal? _Nervous tissue_

11. What part of a plant's vascular system conducts water and minerals from the roots upward?

Xylem

12. Why must a hummingbird eat its own weight in nectar and insects daily?

In order to maintain a high energy level. That allows its heart to beat 1000x/minute.

13. Name the three different types of muscle tissues:

Smooth muscles   *Skeletal muscles*   *Cardiac muscles*

Match the words with their meanings.

14. _g_ Among other roles, this tissue enables an animal to move.

15. _a_ The simplest form of change in an insect.

16. _d_ The site of photosynthesis and food storage in plants.

17. _e_ Partial metamorphosis.

18. _c_ Group of cells that are separated by a "matrix."

19. _h_ After completing the larval stage of holometabolism, the insect becomes this.

20. _i_ Tissue that channels fluids throughout a plant.

21. _f_ Complete metamorphosis.

22. _b_ The arrangement of organs and other parts that allows an organism to survive.

a. ametabolism
b. body plan
c. connective tissue
d. ground tissue
e. hemimetabolism
f. holometabolism
g. muscle tissue
h. pupa
i. vascular tissue

ESSAY

Choose a feature of an animal discussed in this chapter, and explain whether it could, or could not, have evolved by natural selection.

INQUIRY ACTIVITY: *Metamorphosis Lab*

As we discussed in Chapter 10, one of the greatest mysteries in nature is butterfly metamorphosis. After emerging from its egg, a caterpillar must eat as much as it can, grow, and survive in order to begin, and then complete, its metamorphosis into a butterfly. This activity will allow you to watch this process firsthand.

To observe a hungry caterpillar growing into a beautiful butterfly, you will need to find or purchase a caterpillar. The best time of year for this exercise is spring through summer, when there are green plants for food and little chance of rain or snow during the release of the butterfly.

For a quality documentary presenting the evidence for intelligent design from insect metamorphosis, see the film Metamorphosis, available at **www. metamorphosisthefilm.com.**

- **Materials Needed:** A caterpillar or two, leaves from a proper food source; an enclosure; a sturdy twig; ruler; pencil; and paper.

- **Step 1:** Caterpillars need air, and a safe enclosed space. Before you get any caterpillars, you'll need to obtain an enclosure. A good one would be an empty fish tank with a fine mesh cover. If you buy a caterpillar, the seller may also sell good mesh enclosures.

- **Step 2:** Along with the enclosure, you will need a sturdy twig to fit inside it. The twig should be long enough to reach near the top. The pupa may (or may not) use the twig for attaching itself—but you should be hospitable, and offer the twig just in case.

- **Step 3:** Obtain caterpillars. You can obtain caterpillars by either finding them in nature or buying them in a kit. To find one, go to a field and look for leaves with holes or bite-marks— telltale signs of a hungry caterpillar. If you'd rather buy caterpillars, we recommend the website Insect Lore. The site sells Painted Ladies—the most popular type of caterpillar for this kind of experiment and a very easy type to raise. Visit the website at:
http://www.insectlore.com/Living+Kits/Butterflies/

- **Step 4**: At the same time that you do Step 3, obtain leaves from a proper food source. It is often possible to buy food when you buy the caterpillar. But ideally, we recommend picking a fresh source of food. Many caterpillars, including Painted Ladies, eat leaves from plants in the mallow family—a common weed. If you found your caterpillar outdoors, the plant you found it on is most likely its food source. Gather large quantities of leaves and rinse them with water (no soap) before feeding to rinse off any insecticide or uninvited "guests." Collect more food than you think will be necessary—the caterpillar is hungry, after all. Store any extra in a plastic bag in a refrigerator.

- **Step 5:** Once you have your caterpillar and its food, place them inside the enclosure. With the ruler, carefully measure your caterpillar's length, and record it.

- **Step 6:** Every day, measure the caterpillar and make sure it has enough food. Add fresh food daily. As it reaches the end of its growth cycle, it may spend more time at the top of its enclosure or near the twig. Try to avoid disturbing the caterpillar at this time, as it will soon form a pupa. Record any observations you make of the caterpillar each day; you may even choose to take a daily picture of the caterpillar.

- **Step 7:** Eventually the caterpillar will stop eating and attach itself to the roof of the enclosure or the twig. Soon you will see it bend its head and harden into a pupa. Do not shake the container or move the pupa. Continue to make and record observations each day.

- **Step 8:** Inside the pupa, the caterpillar's body is liquefying into a "soup," so the butterfly body plan can form. This process usually takes one to two weeks. When the butterfly is ready to emerge, you will see a reddish-colored liquid dripping from the bottom of the pupa. This will soften the pupa so the butterfly can emerge—either that same day or the next. Continue to make and record observations each day.

- **Step 9:** Do not try to help the butterfly emerge. It will struggle for a while, but it must do this on its own in order to mature properly. If you try to help it (e.g., by cutting the space in the pupa) it will be deformed and unable to fly.

- **Step 10:** After the butterfly emerges, do not touch it until its wings are extended and dry. The wings will look a bit floppy, but after an hour or two, the butterfly will be stronger and will begin flapping its wings. Watch as it experiences being a butterfly. Although it is not required to feed the butterfly, you may choose to place a piece of an orange or a shallow bowl of sugar-water inside the enclosure.

- **Step 11:** Make observations of the fully formed butterfly. Review your observations as the caterpillar grew and then changed into a butterfly. Create a chart showing the daily growth of the caterpillar. How did it change over time? How is the caterpillar stage different from the butterfly? What changes had to take place during the process of metamorphosis?

- **Step 12:** Release the butterfly into the wild. When you are ready to say "goodbye" to your butterfly and release it, take it outside, open the enclosure, and gently allow it to climb onto your hand. Do not grasp its wings, as they can easily be damaged. Lift your hand and let the butterfly fly away.

CHAPTER 11

BODY OF EVIDENCE

12/5/16 ✓

1. Name four necessary components of the vertebrate eye:

- a cornea
- an iris
- ✓ Capillaries (blood)
- Muscles

2. A fully functional eye is _____Useless_____ to an organism if there is no nervous system to process the visual signal and trigger an appropriate behavioral response.

3. What is asexual reproduction?

Asexual reproduction is where an organism clones itself and has no need for a male or female.

4. Under asexual reproduction, an organism passes __100__ % of its genes on to the next generation. But in sexual reproduction, only __50__ % of an organism's genes are to passed each offspring.

5. Fertilization in humans typically takes place in the:

Fallopian tubes

6. Name five specialized features that must be present for sexual reproduction to occur in humans:

- Ovaries
- Unique hormones
- a vaginal entrance

- _a placenta_
- _Amniotic fluid_

It is not compatible with the other ←

7. Give three reasons why sexual reproduction is less favored than asexual reproduction:

- _In sexual reproduction a member of one sex cannot reproduce if_
- _Sexual reproduction cuts fitness in half._
- _Sexually reproducing organisms must take great amounts of energy to create, preserve, and maintain sex cells._

8. The ___villi___ in our small intestines absorbs nutrients.

9. Name five necessary components of our digestive tract:

- _Teeth_
- _Strong jaw bones_
- _Saliva_
- _Muscles_
- _Bile_

10. What is another name for the digestive tract?

Alimentary Canal.

11. What essential component lines and protects the entire 30 feet of the alimentary canal?

Mucus membrane

12. Name three other essential systems in the human body:

- _Musculoskeletal_
- _Respiratory_
- _Endocrine_

13. Choose one of the systems of the human body not elaborated upon in the text and list five necessary components found within it (may need further research): *Auditory System*

- Auricle
- Ear bones
- Cochlea
- Wernicke's area
- Auditory Cortex

14. Do you think any of the systems discussed in this chapter might be irreducibly complex? Explain your answer:

Yes, all in fact. If even one tiny component is wrong the system will either fail or not function at 100% of its ability

Fill in the blanks: Vertebrate Eye

Vitreous gel our

Iris

Optic Nerve

Cornia

Macula

Pupil

Fovea

Lens

Retina

Iris

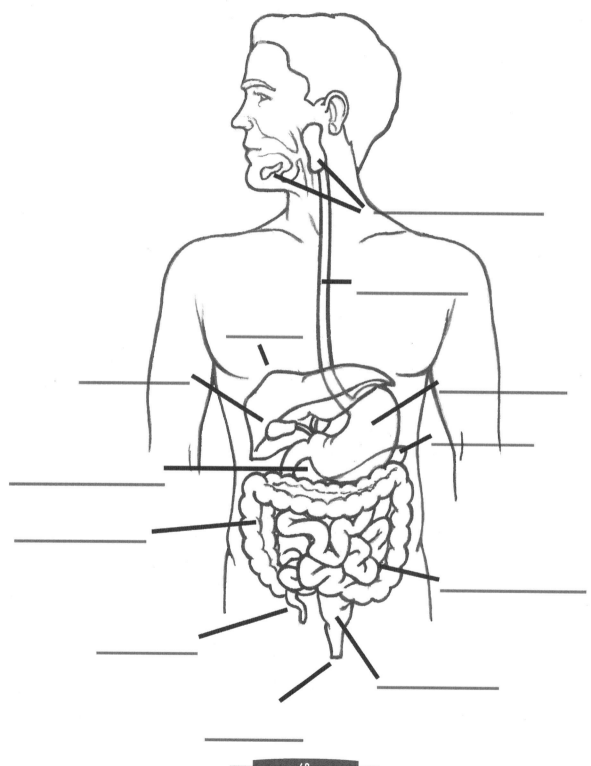

ESSAY

How is the digestive tract like an intelligently designed assembly line?

INQUIRY ACTIVITY: *Eye Exercises*

Vision is an amazing ability involving the intricate, multi-part eye as well as nearby muscles and the nervous system. In these activities, we will explore three abilities: color vision, depth perception, and the ability to focus.

ACTIVITY 1: *Color vision*

- **Time of day:** Late dusk (just before dark).

- **Materials Needed:** A room with a light switch; a few pieces of colored paper (each paper a different color) and a pencil.

- **Step 1:** In the nearly dark room, look at the papers and write on each one its apparent color.

- **Step 2:** Turn on the lights and compare your guesses with the actual colors. Were your guesses correct? Our eyes contain rods and cones, and each has a different function. The rods, while they do not detect color, allow us to see shapes in a relatively dark environment. Conversely, the cones are not effective in dark environments, but they give us color acuity in daylight.

ACTIVITY 2: *Binocular vision / Depth perception*

- **Materials Needed:** A door frame or other straight vertical object; two pencils.

ACTIVITY 2A:

- **Step 1:** Close one eye, hold a pencil vertically at arm's length and line it up with the door frame.

- **Step 2:** Close that eye as you open the other. Notice how the pencil seems to shift relative to the door frame. The human brain compensates for this shift and provides a single image.

ACTIVITY 2B:

- **Step 1:** Hold two pencils, one in each hand.

- **Step 2:** Close one eye and, holding the pencils shoulder high at arm's length, attempt to touch the ends together.

- **Step 3:** Repeat Step 2 with both eyes open.

- **Step 4:** Which method was easier? The relative ease of touching the pencils with both eyes open demonstrates the importance of binocular vision and depth perception in our everyday activities.

ACTIVITY 3: *Focusing ability*

- **Materials Needed:** An object about 20 feet away.

- **Step 1:** Stare at an object about 20 feet away for about one minute.

- **Step 2:** Position one of your fingers near your line of sight of the object, but not in its way.

- **Step 3:** Shift your vision back and forth from the object to your finger. Note how each object quickly comes into focus as you shift your eyes back and forth.

CHAPTER 12

POORLY DESIGNED ARGUMENTS

12/13/16

1. Complete the definition of dysteleology: "The view that nature was not intelligently designed, _often based on the claims that some natural structures are flawed or functionless._"

2. When ID proponents say a structure demonstrates "intelligent design," what do they mean?

They simply seek to indicate that a structure has features requiring a mind capable of forethought to design the blueprint.

3. Does intelligent design require perfect design? Why or why not?

No, because think about it, we intelligently create many things but their designs aren't perfect. Therefore intelligent design does not require perfect design.

4. Arguments that alleged flaws in nature refute ID should be subjected to two steps of critical scrutiny:

- First, _we must determine whether the flaw, if real, would actually be an argument against intelligent design._

- Second, _we must investigate whether the flaw is real._

5. What is the purpose of glial cells in the vertebrate eye?

They channel light through the optic nerve wires directly to the photoreceptor cells.

6. Does the wiring of the vertebrate eye imply a flawed design? Explain your answer:

No, because the vertebrate eye "produces the highest degree of visual quality". Just one slight change could be detrimental to an organism's sight.

7. Some scientists claim the laryngeal nerve has a suboptimal design. Do you agree with them? Why or why not?

I do not agree. ~~Because~~ when people are born with the supposed "optimal" design they suffer from various health issues.

8. The field of _biomimetics_ looks at nature for inspiration when creating human technology.

9. Does the presence of blind eyes in cave-salamanders refute intelligent design? Why or why not? Yes & No. Darwinian Evo. is the best explanation for their blind eyes, however, ID allows that DE can cause loss of function.

It does not because ID accepts that Darwinian processes can accomplish some types of change. ID is more concerned with how complex features like functional eyes are gained in the first place than how they can be lost.

10. What percentage of the human genome do biologists currently think is functional? 80%
List three functions discovered for non-coding DNA:
- Repairing DNA
- Regulating gene expression
- Helping in DNA replication

11. Should the appendix be considered vestigial? Why or why not?

No because it carries beneficial bacteria and some people who've had it removed suffer from health problems.

12. Why is the list of vestigial organs getting smaller?

Because, we are discovering uses for the supposed "vestigial" organs.

13. Given the track record of vestigial organs, should textbooks continue to cite them as evidence for evolution? Why or why not?

No because we are continuously discovering more and more uses for them.

```
appendix          panda's thumb
dysteleology      probiotics
equivocation      pseudogenes
junk DNA          SLN
optic nerve       vestigial
```

Fill in the blanks with the words from the box.

14. _panda's thumb_ An organ used to strip bamboo.

15. _optic nerve_ Part of the vertebrate eye that relays information to the brain.

16. _appendix_ An organ housing beneficial bacteria, which plays a part in the human immune system.

17. _junk DNA_ DNA that many scientists believed was non-functional, until recently.

18. _SLN_ A nerve for the larynx that follows a direct pathway from the brain.

19. _dysteleology_ An argument that nature was not intelligently designed, based on allegedly flawed organs or structures.

20. _vestigial_ A biological structure that supposedly lost its function through evolution.

21. _probiotics_ Beneficial bacteria added to food and drinks.

22. _pseudogenes_ Sections of DNA that materialists claim are broken genes but are increasingly being found to perform functions

23. _equivocation_ Using terms with unclear meanings in order to confuse an opponent or audience.

ESSAY

Of the dysteleological arguments discussed in this chapter—the panda's thumb, the recurrent laryngeal nerve, or the wiring of the vertebrate eye—which one is the best example of a "flawed" design? Is there a rebuttal to that argument? If so, elaborate.

INQUIRY ACTIVITY: _The Shrinking List of Vestigial Organs_

In Chapter 12 we discussed how during the Scopes trial in 1925, an evolutionary biologist suggested that there are over 180 vestigial organs and structures in the human body. But in 2008, the journal _New Scientist_ reported the list "shrank," so that today "biologists are extremely wary of talking about vestigial organs at all."

This chapter reviewed a few organs in the human body that were once thought to be vestigial, but are now known to have a function.

- **Materials Needed:** A computer; the Internet.

- **Step 1:** Using the Internet, identify additional body parts that were once on the "vestigial list," but are now known to perform a function.

- **Step 2:** For each organ, study its function and hypothesize what harm might have been done to patients if Darwinian medical doctors removed it, or if our bodies never had that part to begin it.

CHAPTER 13

TREE HUGGERS

1/4/17 ✓

1. The view that all living organisms are related is called

Universal Common ancestry.

2. What is the main assumption that phylogenetic trees are based on?

The assumption that similarity between organisms is the result of inheritance from a common ancestor.

3. _Morphology_ is the form, structure and body plan of an organism.

4. Define convergent evolution:

Two or more species Independently acquiring the same trait, supposedly through Darwinian evolution.

5. Do you think <u>convergent</u> <u>evolution</u> adequately explains the existence of similar structures in unrelated species? Explain your answer:

No.

Darwinian Evo is supposed to have no goals, however CE implies species are evolving the same complex traits over repeatedly despite the surprising difficulty of doing this even once!

6. Molecule-based trees are constructed by

They are constructed by comparing DNA, RNA, or protein sequences in different organisms.

7. Morphology-based trees are constructed by

They are made by comparing physical characteristics.

8. What is horizontal gene transfer?

It's a process where microorganisms obtain genes through processes other than inheriting them from a parent. (sharing + swapping)

9. Give two reasons why horizontal gene transfer does not validate common descent:

- Though HGT can happen with some bacterial genes, it is very unlikely that it could occur with so many essential genes.

- Among more complex organisms - like animals - where such gene swapping is not directly observed, conflicts with the tree of life become prevalent.

10. Define homology:

Similarity of structure and position, but not necessarily function.

11. What is Berra's Blunder?

It's taking evidence for common design, and mistakingly calling it evidence for common decent.

12. Other than common descent, what is another possible explanation for homology? Why?

Common design is an entirely reasonable explanation for functional anatomical similarities. If the designer creates a functional body plan, why must he start over?

13. _c._ A process by which bacteria can obtain genes through means other than inheritance.

14. _e._ A phylogenetic tree based on comparing similarities in physical characteristics such as anatomical and structural similarities.

15. _f._ Two or more unrelated species evolving similar traits.

16. _b._ Similarity of structure and position, but not necessarily function.

17. _d._ An evolutionary tree based on similarities in DNA, RNA or protein sequences.

18. _g._ A hypothetical tree where "all the organic beings .... have descended from one primordial form."

19. _a._ A hypothesis about the ancestral relationships between organisms.

a. convergent evolution
b. homology
c. horizontal gene transfer
d. molecular tree
e. morphological tree
f. phylogeny
g. universal tree of life

ESSAY

Based on what you've learned in this chapter, which theory offers the best explanation for the similarities between different species: common descent or common design?

INQUIRY ACTIVITY: *Construct a Simple Phylogenetic Tree*

In Chapter 13, we learned that biological similarity can be explained by common design as well as by common descent. This activity will allow you to construct a simple phylogenetic tree to understand the assumptions that go into tree-building, and see whether common descent, or common design, is a better explanation for similarities between organisms.

- **Materials Needed:** Pencil; paper; and three people with pockets full of stuff.

- **Step 1:** Find three people with pockets full of stuff (e.g., cell phone, keys, pens, gum, handkerchief, etc.). Have them empty the contents of their pockets onto a table. Keep each set of contents separate from the others.

- **Step 2:** Imagine that each set is an individual organism, and its characters are the individual items. For example:

  - If a person's pockets contain keys, a cell phone, and a stick of gum, then that "organism" has the following "characters": keys, cell phone, stick of gum.

For the rest of this exercise, an "organism" will refer to a pocket's set of contents, and an item will be called a "character." To avoid confusion, fill out a chart listing the name of the person who owns each "organism." We've given a hypothetical example below.

Chart 1: Name of Person Who Owns Each "Organism"

	Owner's Name
Organism 1:	Sally
Organism 2:	Bob
Organism 3:	Ziggy

- **Step 3:** Draw a phylogenetic tree showing how these "organisms" are most likely related. This is a multistep process:

- **Step 3a:** Looking at all the characters combined, choose three that you will use to construct your tree. For this exercise to work, choose at least one character that is not present in all three organisms.

- **Step 3b:** Make a chart showing which characters are present in each organism. To do this, use a chart like the one below. If the character is present, score a "1." If it's absent, score a "0." We have provided a hypothetical dataset below to help you understand how this works:

Chart 2: Scoring Characters for Organisms 1-3.

Character	Organism 1	Organism 2	Organism 3
Cell phone	1	0	1
Keys	1	0	0
Gum	1	0	1

- **Step 3c:** Next, make another chart to show how many similar characters are found between each of the three organisms. Look at the data in Chart 2 to fill this out correctly:

Chart 3: Which Organisms Have the Most Similarities?

	Number of Similar Characters
Organism 1 vs Organism 2	0 (nothing in common)
Organism 1 vs Organism 3	2 (cell phone and gum)
Organism 2 vs Organism 3	1 (no keys)

- **Step 3d:** Now you will draw a phylogenetic tree based upon the assumption that those organisms with the most similar characters are most closely related. Use the data in Chart 3 to do this. Because there are only three organisms in this exercise, there are only a few different possible ways to organize their relationships. They are as follows:

Tree A: Organisms 1 and 2 are most closely related.

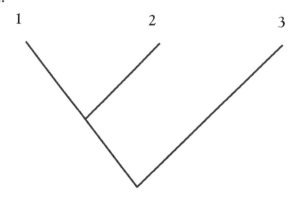

Tree B: Organisms 1 and 3 are most closely related.

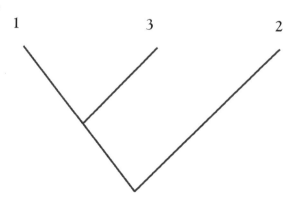

Tree C: Organisms 2 and 3 are most closely related. 2 3 1

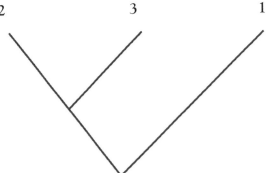

Which tree reflects the most likely relationships between the organisms in this hypothetical dataset? Because Chart 2 showed that organisms 1 and 3 have the highest number of similar characters, they should be grouped as the most closely related. Thus, in this hypothetical dataset, Tree B would be the best choice.

- **Step 4:** Did you encounter any problems when trying to choose the best tree? In some cases, there won't be a single comparison of two organisms that stands alone as having the highest number of similar character. For example, your dataset could possibly have come out like this:

Chart 4: Which Organisms Have the Most Similarities?

	Number of Similar Characters
Organism 1 vs Organism 2	1
Organism 1 vs Organism 3	1
Organism 2 vs Organism 3	1

In this chart, which comparison shows the highest number of similar characters? The answer is none—they all share one character in common, and no two organisms stand out as most similar. In this case, you have encountered one of the problems that evolutionary biologists commonly face: the data do not fit a tree-like pattern.

- **Step 5:** Even if you were not able to fit the data to a tree, assume for a moment that you were. Consider the following questions:

  - What assumptions did you make in building the tree?

- Does the tree reflect *actual* common ancestry?

- If not, does the fact that you can make assumptions, and build a hypothetical tree, necessitate that common descent is true?

- What other explanations might there be for the shared similarities (and differences) between the "organisms" you used in this exercise?

CHAPTER 14

FAKES AND MISTAKES

1/10/17

1. What does the phrase "ontogeny recapitulates phylogeny" mean?

It refers to the recapitulation theory. Define it

Why is this idea false? The evidence shows that vertebrate embryos do not replay their supposed earlier evolutionary stages.

2. Name the biologist who made the drawings called "one of the most famous fakes in biology":

Haeckel

3. What did he draw, and what was the problem with the drawings?

He drew embryos, but they did not resemble the way actual living embryos look, plus he didn't even draw them at their earliest stages like he proclaimed.

4. Name the theory that acknowledges that vertebrate embryos start their development differently but still postulates that they pass through a highly similar "pharyngula" stage midway through development:

Developmental hourglass theory.

5. List three ways actual vertebrate embryos differ at the pharyngula stage:

- Body size
- Body plan
- Growth patterns

6. Define adaptive radiation:

The supposed rapid diversification of species after entering an empty habitat, or niche.

7. Are the Galápagos finches an impressive example of evolution? Why or why not?

No, because the finches don't suddenly become different species because of their beak structure. They are just a natural example of microevolution.

8. What are two weaknesses of the peppered moth story?

- _The change in moth coloration is an example of microevolution; not macroevolution._
- _Many photographs in textbooks showing the moths on tree trunks are staged._

Match the words with their meanings.

9. _g._ The development of an organism.

10. _d._ Fraudulent portrayal of vertebrate development that obscures the differences between embryos.

11. _c._ An exaggerated feature in embryo drawings, used to imply humans are related to fish.

12. _i._ Means "replays" or "repeats."

13. _h._ The allegedly similar mid-point in vertebrate embryo development.

14. _e._ An organism's theoretical evolutionary history.

a. adaptive radiation
b. drought
c. gill slits
d. Haeckel's embryo drawings
e. ontogeny
f. peppered mouth
g. pharyngula stage
h. phylogeny
i. recapitulates
j. Jonathan Wells

15. __f.__   An insect in England that has oscillated between light and dark colored forms.

16. __b.__   Galápagos finch beak sizes returned to normal after this ended.

17. __a.__   The idea that organisms diversify quickly when entering into an empty niche.

18. __j__   Biologist who wrote *Icons of Evolution*.

ESSAY

Give an example of the evolution "bait and switch" tactic. Explain whether it is persuasive as an argument for Darwinian evolution.

INQUIRY ACTIVITY: *Exploring Vertebrate Embryo Similarities and Differences*

In Chapter 14, we learned that biology textbooks often overstate the degree of similarity between vertebrate embryos in their earliest stages. This activity will allow students to compare vertebrate embryo drawings often found in biology textbooks to actual photographs of vertebrate embryos in their early stages of development.

- **Materials Needed:** Haeckel's original embryo drawings; scissors; and photographs or accurate drawings of actual embryos such as those found in:

   (1) Jonathan Wells, "Haeckel's Embryos and Evolution: Setting the Record Straight," *The American Biology Teacher*, 61 (May, 1999): 345-349.
   Note: This article is available free online at **http://www.discovery.org/f/629**

   (2) Michael K. Richardson *et al.*, "Haeckel, Embryos, and Evolution," *Science*, 280 (May 15, 1998): 983-985.
   Note: This article is probably available free online by searching for its title, "Haeckel, Embryos, and Evolution."

   (3) Michael K. Richardson *et al.*, "There is no highly conserved embryonic stage in the vertebrates: implications for current theories of evolution and development," *Anatomy and Development*, 196 (1997): 91-106.
   Note: This article is probably available free online by searching for its title, "There is no

highly conserved embryonic stage in the vertebrates: implications for current theories of evolution and development."

(4) Elizabeth Pennisi, "Haeckel's Embryos: Fraud Rediscovered," *Science*, 277 (September 5, 1997): 1435a.
Note: This article is probably available free online by searching for its title, "Haeckel's Embryos: Fraud Rediscovered."

- **Step 1:** Download reference (1) above, and print out Haeckel's drawings from Figure 1. Look at Haeckel's embryo drawings from Chapter 14. What species are portrayed? Cut out the drawings for each organism and organize them according to species.

- **Step 2:** Download references (2)-(4) above and print out photographs / accurate drawings of vertebrate embryos from the articles you collected. Cut out the drawings for each organism and group them according to species with the other organisms from Haeckel's drawings you already cut out. Paste drawings or photographs of the same species next to each other.

- **Step 3:** Compare and contrast Haeckel's embryo drawings with the photographs / accurate drawings. Answer the following questions:

  - What are the similarities?

  - What are the differences?

  - Why do you think some authors called Haeckel's drawings "fraudulent"? Should these drawings be used in textbooks?

CHAPTER 15

SUDDEN, GRADUAL CHANGE

1/22/17 ✓

CHAPTER 15: SUDDEN, GRADUAL CHANGE

1. Name four explosions mentioned in this chapter where new life forms appear in the fossil record without evolutionary precursors:

- Big Bang Bloom of angiosperms
- Formation of our solar system fish, genus Homo
- Earth becomes habitable birds, mammals
- Cambrian explosion

2. Identify a biological structure that would have to arise rapidly during one of the explosions of life on Earth:

sp Mamal explosion

Could Darwinian processes account for its rapid appearance? Explain:

I do not believe so because there are too many gaps in the fossil record and no transitional fossils to fill them.

3. What was Darwin's explanation for the lack of transitional fossils?

"The extreme imperfection of the geological record."

4. Today, what do paleontologists call Darwin's explanation for the lack of intermediate fossils?

Artifact hypothesis.

5. Do you find this explanation persuasive? Why or why not?

No I do not, because studies have shown that this is no longer sp aplicable.

6. What is the theory of punctuated equilibrium (punc eq)?

Most evolution takes place in small populations over relatively short geological time periods.

7. What is the basic problem with the punctuated equilibrium model?

Might require too much genetic change too fast.

8. Define evo-devo:

Theory that states that changes to the master genes that control the development of an organism can cause big, abrupt changes in body plans.

9. Of the problems with evo-devo mentioned in the text, which do you think is the most severe?

The best examples of evolutionary change produced by this theory are meager and often entail loss, not gain.

Why?

Because without specifically good results the theory won't survive.

10. Below are three potential models of the history of life. Which one do you think is best supported by the fossil evidence?

The Explosions one.

Model 1: Gradualism

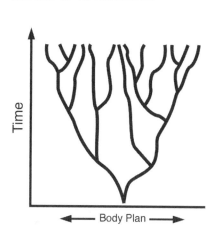

Time

← Body Plan →

Model 2: Punc Eq

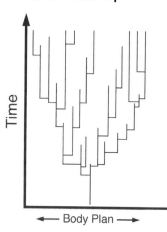

Time

← Body Plan →

Model 3: Explosions

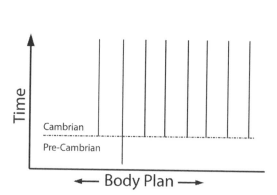

Time

Cambrian

Pre-Cambrian

← Body Plan →

Why did you choose this model?

Because this sums to make the most sence with the evidence provided, and my religion could explain the sudden rise of fossils.

Fill in the blanks with the words from the box.

~~artifact~~ hypothesis   ~~explosion~~
~~big bloom~~   ~~fossils~~
~~Cambrian explosion~~   ~~Hox~~ genes
~~evo-devo~~   ~~punc eq~~

11. _punc eq_ Model where evolution takes place rapidly in small populations.

12. _evo-devo_ Combination of two fields that claims that changes to developmental genes cause large, abrupt changes in body plans.

13. _big bloom_ An explosion in the fossil record showing the abrupt appearance of flowering plants.

14. _Cambrian Explosion_ An event in the fossil record where most animal phyla appear abruptly.

15. _artifact_ The claim that species appear abruptly in the fossil record because
    hypothesis it is imperfect.

16. _Hox genes_ Changes to these are usually deadly, but are nonetheless theorized to
    explain the rapid appearance of new body plans.

17. _explosion_ A common phenomenon in the fossil record where organisms appear
    abruptly, without evolutionary precursors.

18. _fossils_ The preserved remains of dead organisms.

ESSAY

Is intelligent design a logical explanation for the abrupt appearance of new body plans in the fossil record? Why or why not?

INQUIRY ACTIVITY: _Making a Collectors' Curve_

Materialists often try to explain the lack of transitional fossils by arguing that the record is incomplete. Some paleontologists have responded by pointing out that our knowledge of the fossil record may be imperfect, but it is fairly complete—meaning that transitional fossils are in fact conspicuously lacking.

How did paleontologists conclude that the fossil record is relatively complete? They did this, in part, by constructing collectors' curves—a type of graph used by scientists in many fields to measure the completeness of a collection. In this inquiry activity, you will construct a mock collectors' curve and assess the completeness of your collection. Our "specimens" will be pocket contents.

- **Materials Needed:** Graph paper; pencil; and at least three friends with pockets (or purses) full of stuff. The more friends you find, the better this exercise will work.

- **Step 1:** Using graph paper, prepare your collectors' curve graph with an X-axis and a Y-axis. The X-axis should be labeled "Total Number of Specimens Found," and should be along the long (horizontal) edge of the paper. The Y-axis should be along the short (vertical) edge of the paper, and should be labeled "Total Types of Specimens Found." It should look something like this:

Total Types of Specimens Found

Total Number of Specimens Found

- **Step 2:** Prepare a data table like the one below to keep track of the specimens. The table should have three columns: Specimen Number, Specimen Type, and Graph Coordinates. Each specimen will require its own row.

Specimen Number	Specimen Type	Graph Coordinates

- **Step 3:** Find at least three friends with stuff in their pockets. Each item in their pockets will be called a "specimen."

- **Step 4:** Collect specimens from pockets, and record them in the first two columns in the data table. Be sure to record the specimens in the order you discover them. A sample of what the table might look like is seen below:

Specimen Number	Specimen Type	Graph Coordinates
1	Keychain	
2	Cell phone	
3	Keychain	
4	Wallet	

- **Step 5:** Fill out the third column by determining the graph coordinates. The X-axis value will increase by one for every specimen, but the Y-axis value will only increase when you discover a new type of specimen. Our sample table has been filled in below, with an explanation added as a fourth column:

Specimen Number	Specimen Type	Graph Coordinates	Explanation
1	Keychain	(1, 1)	For the first specimen, the X-axis value will be 1. The first specimen is also the first keychain, meaning that its Y-axis value will also be 1. Its coordinates are (1, 1).
2	Cell phone	(2, 2)	The second specimen will have an X-axis value of 2. Since it is also the second type of specimen, it has a Y-axis value of 2. Its coordinates are (2, 2).
3	Keychain	(3, 2)	The third specimen has an X-axis value of 3. But, at this point you've still only found 2 total types of specimens—keychain and cell phone—so its Y-axis value is 2. Its coordinates are (3, 2).
4	Wallet	(4, 3)	The fourth specimen is a wallet. This is the fourth specimen, so its X-value is 4. It's also a new type of specimen—the third type you've found. So its Y-value is 3. Its coordinates are (4, 3)

- **Step 6:** Once you've determined the graph coordinates for each specimen, it's time to create your collectors' curve. Plot each coordinate in the data table on your graph, then draw a line connecting the dots. It might look something like this:

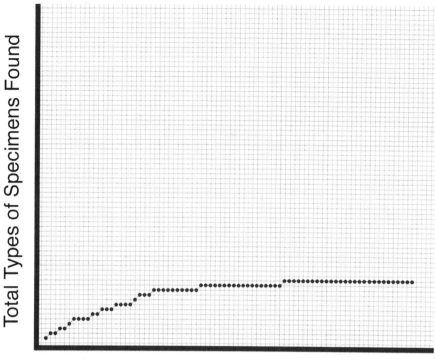

Total Number of Specimens Found

As you surveyed more and more specimens, you probably eventually found that each new specimen surveyed was of a type you'd already seen. At this point, your collectors' curve has "leveled off," and has a slope close to zero. Does your curve "level off"?

When the curve begins to level off, it means you have a good idea that your knowledge of the different types of specimens is relatively complete. In the same way, when paleontologists see that collectors' curves of fossils "level off," they believe they have a fairly complete knowledge of the different types of fossil species.

CHAPTER 16

SKELETONS IN THEIR CLOSET

2/8/17

CHAPTER 16: SKELETONS IN THEIR CLOSET

1. What is a theropod?

A type of carnivorous and bipedal dinosaur with small forelimbs.

2. What living animals do materialists claim some theropods evolved into?

Birds.

3. Why did *Archaeopteryx* excite many scientists initially?

Because it was said to be an intermediate form between birds and theropods. It also had feathers like a bird.

4. Proponents of the reptile-to-bird hypothesis must explain transitions between some major differences in the two classes. List four of those changes:

- Cold-blooded   > Warm-blooded
- Slow metabolism   > Fast metabolism
- No feathers   > feathers
- Three-chambered heart > 4 chambered heart

5. Which do you think would have been the most difficult change for Darwinian evolution to produce? Why?

The heart change because you'd have to somehow still have blood flowing through the heart while it was changing.

6. The problem with most alleged feathered dinosaur fossils is that they either aren't

___feathered___ , aren't ___dinosaurs___ , or aren't ___fossils___ .

7. Complete the timeline:

Fossil	Conventional Age (in millions of years)
Tetrapod tracks (reported in 2010)	397
Hypothetical fish ancestors of tetrapods	390
Tiktaalik roseae	375
Tetrapod body fossils	360

8. How do the dates in the chart above contradict Neil Shubin's claim that *Tiktaalik* was found in "rocks of just the right age"?

___Tetrapods were found before Tiktaalik, way before___
___(20 million yrs before)___

9. List two weaknesses in how horse fossils are used to support the theory of evolution:

• ___Some earlier horses in the fossil record are larger than later ones.___

• ___Fossils portrayed in the "lineage" span over different continents.___

10. Of all the required changes between land mammals and whales, which one do you consider to be the most difficult for evolution to produce?

___The emergence of a blowhole.___

Explain why: ___When it is developing won't the creature have a hard time breathing?___

11. What makes whale fossils a case study for problems with evolution?

Because it has so many age differences between the fossils it pokes holes in evolution. (short amt. of time allowed by the fossil record)

Match the words with their meanings.

12. _d._   A fossil initially promoted as an ancestor of humans.

13. _b._   A fossil alleged to be an intermediate between reptiles and birds.

14. _e._   Birds that lost their ability to fly.

15. _g._   Carnivorous, bipedal dinosaurs with small forelimbs.

16. _a._   Unique breathing structures in birds

17. _c._   Wispy, hair-like structures that are not feathers.

18. _h._   A fossil alleged to be an intermediate between fish and tetrapods.

19. _f._   Four-legged vertebrate.

a. air sacs
b. Archaeopteryx
c. dinofuzz
d. Ida
e. secondarily flightless
f. tetrapod
g. theropod
h. Tiktaalik

ESSAY

Based on what you've learned from Chapters 13-16, has your opinion changed about the validity of universal common descent? Explain.

INQUIRY ACTIVITY: *Investigating Feathers*

In this activity, we will explore some of the well-designed features of feathers: the barbs, barbules, and shaft. This will help demonstrate the irreducibly complex structure of feathers.

ACTIVITY 1: Barbs and barbules

- **Material Needed:** A feather and a pencil.

- **Step 1:** Hold the shaft of the feather with one hand. With the other hand, gently grasp one side of the feather between two fingers, and stroke it from the top down. The barbs and barbules will now be separated. Note the rough appearance of the feather.

- **Step 2:** Using the same motion, stroke the feather in an upward motion. This mimics a bird's preening action. Note that the barbs and barbules have now been "zipped" back into place and the feather is in position for flight.

- **Step 3:** With your fingernail or the point of a pencil, gently attempt to separate two of the barbs. Note that the barbs and barbules tend to cling together.

ACTIVITY 2: Shaft (a.k.a. rachis)

- **Materials needed:** A 6" x 6" square of cardstock, and a piece of tape.

- **Step 1:** Holding the cardstock with one hand, gently bend it with a finger of the other hand. Note how easy it is to bend.

- **Step 2:** Twist the cardstock into a tight cylinder and tape it in that shape. This represents the hollow structure of the feather shaft.

- **Step 3:** Again using a single finger, try to bend the twisted cardstock. Note how rigid it is. This rigid structure is one of many aspects of feathers that combine to enable flight.

CHAPTER 17

WHO'S YOUR DADDY?

2/26/17

1. What is paleoanthropology? *The study of the origin of humans.*

2. Why are biases and bitter disputes so common in that field?

Because it isn't based on objective criteria, but instead on subjective interpretations of little fossil evidence.

3. What organisms are included within hominins?

They include humans, chimpanzees, and any extinct organisms back to their presumed most recent common ancestor.

For questions 4-13, match the description with the best answer listed in the box below. Some answers may be reused multiple times. Multiple answers to a question are possible.

| a. Australopithecus | b. Homo | c. Neanderthals |
| d. Homo sapiens | e. Homo erectus | f. habilis |

X4. *a* *e.* Capable of some upright walking; long forelimbs indicate they spent significant time in trees.

5. *C.* Very likely a sub-race of *Homo sapiens*.

6. *a.* Had inner ear canals similar to those of great apes.

X7. *f d.* Dated at about 1.9 mya, but does not predate the earliest true members of *Homo*, and thus could not have been a precursor of our genus.

8. *b.* Separated from australopithecines by a large morphological gap.

9. *a.* Had the hands of knuckle-walking apes.

10. *C.* Had an average skull size greater than that of modern humans, and anatomical evidence suggests they were capable of speech.

X11. *e a.* Had an average skull size smaller than modern humans, though within the range of modern human variation; virtually identical to modern humans below the neck.

12. ___f.___ More similar to modern apes than were other australopithecines.

13. ___b d.___ Modern humans are members of this species.

14. Does the fossil evidence support the idea that Lucy was an ancestor to humans? Why or why not?

No because Lucy bears more resemblance to apes than humans and only 40% of her bones were found.

15. Describe two of the weaknesses in the claim that human and chimp DNA are only 1% different:

• Genetic comparisons usually only focus on differences in protein-coding DNA and not "junk" DNA.

• In some ways our genome is very different from that of chimps. - such as the "y" chromosome.

16. List three human behaviors that challenge natural selection:

• Always walk upright more celibacy, martyrdom

• relatively hairless voluntary poverty, altruism, building

• Writes poetry, more like this   Cathedrals
                                                        etc.

17. Pick one behavior listed above and explain why it challenges Darwinian evolution:

Being mostly hairless poses a challenge in my mind because if Darwinian evolution is for the better you'd think we'd keep hair all over our bodies for protection and warmth.

I'll grant you this?! :)

Fill in the blanks with words from the box.

```
Australopithecus      kin selection
habilis               Neanderthals
Hominidae             paleoanthropologist
Homo                  reciprocal altruism
human exceptionalism
```

18. A term describing the unique and special qualities of our species:

human exceptionalism

19. An ape-like genus often claimed to be ancestral to humans:

Australopithecus

20. Helping a sibling survive and reproduce:

Kin selection

21. An act of generosity with the hope of a return favor:

Reciprocal altruism

22. A sub-group of the genus *Homo* often portrayed as unintelligent brutes, though they likely possessed language, art, and culture:

Neanderthals

23. A scientist who studies human origins:

Paleoanthropologist

24. A species claimed to be an intermediate link between *Australopithecus* and humans, but later realized to be even more ape-like than *Australopithecus*:

Habilis

25. The genus that includes humans:

Hominidae

26. The primate family that includes humans and australopithecines:

Homo

ESSAY

Are humans just another animal, or do we have traits that distinguish us from apes, making humans exceptional? Give examples and explain your answer.

INQUIRY ACTIVITY: *Exploring Altruism vs. Truly Selfless Behavior*

In Chapter 17, we discussed how Darwinian evolution claims that behavior that appears to be altruistic actually gives some self-benefit. Under that view, true selflessness, or pure altruism, doesn't exist. Some people would cynically agree that all human behavior is ultimately selfish. But we think that if anything deserves cynicism, it's Darwinian evolution, which denies the possibility—and the reality—that truly selfless acts commonly take place.

In this inquiry activity, you will interview people who have embarked on different career paths. Your purpose will be to investigate whether they chose their professions based on selfless or ultimately selfish motives.

- **Materials Needed:** Five adults: A teacher, a business professional, a scientist or doctor, a member of the clergy, and a stay-at-home parent, each of whom is willing to have an honest and frank conversation about why they chose their career. This activity can work well even if you don't know adults who have chosen those five specific career paths. But try to find five adults who have chosen diverse careers. You'll also need an audio recorder.

- **Step 1:** Write down a list of questions to ask each of the five adults in order to investigate whether they chose their respective career paths for selfish or selfless reasons. Some suggested questions include:

  - Why did you choose the career you did?
  - Do you serve other people in your job? If so, do you enjoy it?

- Do other people serve you in your job?
- What personal benefits do you get from your job?
- Were there other more lucrative career paths you could have chosen? If so, why didn't you choose them?
- Do you think you chose your job out of a desire to serve others, or out of a desire for personal gain?
- Were your motives a mixture of selfish and selfless ones?

- **Step 2:** Using the list of questions you prepared in Step 1, interview each of the five people—the teacher, the business professional, the scientist or doctor, the member of the clergy, and the parent—in separate interviews. For consistency, ask each person the same questions. Record each interview.

- **Step 3:** Analyze the answers given by each person. Consider the following questions for each:
  a. Do you think their motives for choosing their career path were selfish or selfless?
  b. Is it possible they had a mix of motives?
  c. Even if there were selfish motives, were there any truly altruistic motives in addition? List those motives.
  d. If your interview subject had a mix of selfish and selfless motives, does that negate the possibility that at least some of the person's motives were truly altruistic?
  e. In a Darwinian context, would you expect there to be any truly altruistic motives at all? Why or why not?

CHAPTER 18

TAKING INVENTORY

CHAPTER 18: TAKING INVENTORY

1. The first tenet of materialism contends that:

    Either the universe is infinitely old, or it appeared by chance, without cause.

    Do you think the evidence supports this? __No__ Explain your answer:

    The universe had a beginning and isn't infinitely old, because the Big Bang theory requires that it had a beginning and much evidence supports it.

2. The second tenet of materialism contends that:

    The physical laws and constants of the universe were produced by purposeless, chance processes.

    Do you think the evidence supports this? __No__ Explain your answer:

    The laws are finely tuned for complex life so some complex planning had to go into it.

3. The third tenet of materialism contends that:

    Life originated from inorganic material through blind, chance-based processes.

    Do you think the evidence supports this? __No__ Explain your answer:

    Life is so complex that chance could never have enough time to spit out a form of life.

4. The fourth tenet of materialism contends that:

    The information in life arose by unguided, blind processes.

    Do you think the evidence supports this? _No_ Explain your answer:

    The evidence supports that DNA, an essential piece for life, is too complex to be formed blindly.

5. The fifth tenet of materialism contends that:

    Complex cellular machines and new genetic features developed over time through purposeless, blind processes.

    Do you think the evidence supports this? _No_ Explain your answer:

    through blind processes

    Acid sequences that are stable are to rare to produce new genetic features.

6. The sixth tenet of materialism contends that:

    All species evolved by unguided natural selection acting upon random mutations.

    Do you think the evidence supports this? _No_ Explain your answer:

Just random mutations wont evolve a species; many complex pieces have to be formed and put together.

7. The seventh tenet of materialism contends that:

   All living organisms are related through universal common ancestry.

   Do you think the evidence supports this? __No__ Explain your answer:

   Evidence has proven that this is flat out false.

~~~~~~

ESSAY

Explain the positive case for design as you would to a younger sibling or friend. Do you find the explanation convincing?

INQUIRY ACTIVITY: *Random vs. Intelligent Search*

In Chapter 18, we compared intelligent design to a "goal directed" process, and Darwinian evolution to a search that is a "blind, trial-and-error" process. Some websites make it possible to test the difference between a "blind" search and an intelligently directed one.

- **Materials Needed**: Computer and Internet connection.

- **Step 1:** Write down three topics on which you want to find information.

- **Step 2:** Go to the Wikipedia home page (**http://en.wikipedia.org/wiki/Main_Page**). Search Wikipedia for information on those topics using the site's intelligent search engine. You can do this by using the Search box in the upper-right-hand corner of the screen. Type each of the three topics into Wikipedia's intelligent search box. What information was returned? Was it relevant to what you were looking for?

- **Step 3:** Now try a random search. You can do this by going to Wikipedia's homepage (**http://en.wikipedia.org/wiki/Main_Page**) and looking at the links on the far left-hand-side of the screen, finding the "Random article" link. For each of the three topics you chose, click the "Random article" link ten times. How did Wikipedia do? Did it ever return a page that had anything to do with the information you were seeking?

- **Step 4:** Compare and contrast the results of your random and blind search vs. an intelligently programmed one. How did the searches do? What are the implications for intelligent design vs. Darwinian evolution?

CHAPTER 19

MATERIALISM OF THE GAPS

4/7/17

1. Define the fallacies of reasoning below:

• Straw Man: Misrepresenting an opponent's argument, and then attacking the false version.

• Professional Intimidation: Citing one's educational training or professional accomplishments to silence another's viewpoint.

• Genetic Fallacy: Attacking the origin of an argument rather than the argument itself.

• Argument from Popularity: Urging people to believe an idea because it is popular.

• False Choice: Portraying two options as mutually exclusive when they really aren't.

2. In the three stages of truth as formulated by Arthur Schopenhauer, what happens first?

It is ridiculed.

What is the last stage? It is accepted as being self-evident.

3. Name the heavily fictionalized play from 1955 that attempts to portray evolutionists as rational and opponents of Darwin as ignorant bigots: Inherit The Wind.

4. Name the teacher who, in Tennessee in 1925, was convicted of teaching human evolution?

John Scopes.

5. What 2008 documentary featured scientists and scholars who had suffered damage to their careers for daring to support ID? Expelled: No Intelligence Allowed.

6. Which group supports exposing students to different scientific views about Darwinian evolution?

ID Movement.

Which group does not support such exposure? _Darwin Lobby_

7. By percentage, what portion of the voting public agrees that we should teach students about the scientific evidence both for and against evolution?

_78%_

8. Define "God-of-the-gaps" thinking:

I believe it is where people insert God into gaps in science to explain them.

9. Do you think intelligent design reasoning fits a "God-of-the-gaps" model? _No_

Explain your answer: _It only suggests the idea of an intelligent being, who created the universe and it flat out rejects gaps-based reasoning._

Match the words with their meanings.

10. _j._ The reluctance to accept new ideas.

11. _c._ Said, "All truth passes through three stages."

12. _a._ First checkpoint Darwin lobbyists use to censor pro-ID information.

13. _b._ A personal attack to discredit an opponent.

14. _h._ Carefully selects sources of information to be broadcast to the public.

15. _d._ Using outlandish rhetoric to play on emotions and force someone into agreement.

a. academy
b. *ad hominem* attack
c. Arthur Schopenhauer
d. bully tactics
e. fallacy of reasoning
f. *Inherit the Wind* stereotype
g. materialism
h. media
i. NCSE
j. staying in the comfort zone
k. unwarranted conclusion

16. __k.__ Over-extrapolation of the evidence to make an unjustified argument.

17. __g.__ A leading activist group in the Darwin Lobby.

18. __e.__ A debate tactic used when the facts are not in your favor.

19. __f.__ Tendency to portray evolutionists as intelligent, articulate, and open-minded, and opponents of Darwin as ignorant bigots.

20. __l.__ Richard Lewontin said all scientists must maintain a commitment to this.

ESSAY

If you "follow the evidence wherever it leads," where does it go—toward materialism or intelligent design? Elaborate on the evidence that led you to reach that decision.

INQUIRY ACTIVITY: *Intelligent Design and Darwinian Evolution in the Media*

Every theory, regardless of whether it is accepted or rejected, deserves to be represented as it is defined and described by its proponents. But is intelligent design accurately represented in the media? Is the media objective when it comes to covering topics like evolution?

- **Materials Needed:** News-media sources (newspaper, magazines, journals, TV, Internet articles).

- **Step 1:** Now that you have learned how intelligent-design proponents define and describe their theory, find three examples in the media where ID is discussed. Is it accurately represented? Is there a bias evident in the articles you considered? Are any stereotypes used? Be sure to note the source, quote any misrepresentations, and provide a critique.

- **Step 2:** The mainstream media commonly defend Darwinian evolution—but they often use logical fallacies to do so. Find three examples where the mainstream media discuss Darwinian theory. Is it represented with a positive or negative bias? Are any logical fallacies used to defend Darwin's theory? Be sure to note the source, explain any logical fallacies, and provide a critique.

CHAPTER 20

ANSWERING
THE CRITICS

1. What are the four usual steps of the scientific method?

- Observation
- Hypothesis
- Experimentation
- Conclusion

2. Do you think it's a good process? _Yes_ Why or why not?

If followed it is an efficient and easy method to follow that shouldn't be easily convaluded.

3. What additional steps have been added to the scientific method?

- Falsifiable predictions
- Peer review

Do you agree with these added steps? Explain:

Yes because it can help in some ways that the other steps can't.

4. Define peer-review:

A process by which scientists double-check the work of their peers to make sure it has been done right.

5. List two predictions of intelligent design:

- Genes and other functional parts will be commonly reused in different organisms

- Much so-called "junk DNA" will turn out to perform valuable functions.

Do you think the evidence confirms these predictions? Why or why not?

Yes because the evidence makes a good point and it has some sense to it, unlike some.

6. Do you agree with materialists that ID is not scientific, is religious, or is just politics? __No__

Elaborate on your answer:

ID supporters say themselves it's a science, they do their best to not associate themselves with any certain religion, & they aren't trying to appeal to the politics, just the evidence.

7. What is the name of the court case in which a federal judge banned intelligent design from a school district in Pennsylvania?

Kitzmiller v. Dover

List two problems with the court's ruling:

- Attempting to turn science into a voting contest.

- Adopting an unfair double standard of legal analysis.

8. What is the goal of the Darwin Lobby?

Seeks to restrict free speech & intellectual inquiry.

9. What is the goal of the ID movement?

Seeks to expand intellectual inquiry and free-speech rights.

10. What is the only thing that matters in science?

The search for truth.

Why is it the only thing that matters? _Because truth, no matter how absurd, will slowly help us better understand this world in which we live._

11. List two practical steps you can take to stay informed about the debate in the future:

- _Regularly visiting Evolution News & Views._
- _Attend Discovery Institute's Summer Seminars on intelligent design._

For questions 12-17, match the quotation with its author. Each answer is used only once.

a. Stephen Jay Gould	b. Judge John E. Jones	c. Kenneth Miller
d. Steven Weinberg	e. Jonathan Wells	f. Jay Wexler

12. ____a.____ "The quality of a scientific approach or opinion depends on the strength of its factual premises and on the depth and consistency of its reasoning, not on its appearance in a particular journal or on its popularity among other scientists."

13. ____d.____ "[T]he teaching of modern science is corrosive of religious belief, and I'm all for that!"

14. ____b.____ "ID is an interesting theological argument, but ... not science."

15. ____c.____ ID "is always negative, and it basically says, if evolution is incorrect, the answer must be design."

16. ____f.____ "The part of *Kitzmiller* that finds ID not to be science is unnecessary, unconvincing, not particularly suited to the judicial role, and even perhaps dangerous both to science and to freedom of religion."

17. ___c.___ "The outcome of this scientific revolution will be decided by young people who have the courage to question dogmatism and follow the evidence wherever it leads."

18. Fill in the blanks in the following quotation by William Dembski:

"Our critics have, in effect, adopted a _zero- concession policy_
toward intelligent design. According to this policy, _absolutely nothing_
is to be conceded to intelligent design and its proponents. This is especially difficult for novices
to accept.... The point is not to induce a cognitive shift in our critics, but instead to _clarify_
our arguments
to address weaknesses in our own position, to identify areas requiring further work and study,
and, perhaps most significantly, to _appeal to the undecided middle_
that is watching this debate and trying to sort through the issues."

Match the words with their meanings.

19. ___a.___ Laboratory that studies ID.

20. ___c.___ Signer of the Third Humanist Manifesto.

21. ___b.___ The logical end result.

22. ___h.___ A process used to test ideas and reach scientifically valid conclusions.

23. ___f.___ Double-checking scientific results by fellow scientists.

24. ___d.___ Extracurricular student groups that promote ID on school campuses

25. ___g.___ Irrelevant to settling scientific debates.

26. ___e.___ A statement about something you have noticed.

a. Biologic Institute
b. conclusion
c. Eugenie Scott
d. IDEA Clubs
e. observation
f. peer-review
g. religious beliefs and motives
h. scientific method

ESSAY

Of the four examples given in Chapter 20 showing how ID uses the scientific method to make a positive case, which do you think is the strongest? Explain your answer.

INQUIRY ACTIVITY: *Taking the Next Steps*

If you are so inclined, what steps would you take to defend academic freedom for intelligent design? Write an action plan describing how you can promote and defend ID in your school or local community.

- **Materials Needed:** Creativity; inspiration; perspiration; and imagination.

- **Step 1:** Brainstorm three ways you can promote ID in your community. Explain how you think people will respond.

- **Step 2:** If you advocate ID, it's likely that some people will disagree. If people get emotional or upset at you, how will you respond? Write a hypothetical dialogue between yourself and a hypothetical person who is angry with you for defending ID. How should you respond? Through the dialogue, consider how being respectful might help your opponent become willing to listen to your viewpoint. As you write the dialogue, consider this question: Does being respectful to an opponent necessarily mean you have to agree with them?

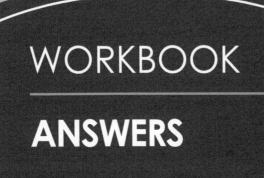

WORKBOOK

ANSWERS

WORKBOOK ANSWERS

Chapter 1

1. intelligent design

2. *Answers may vary.*

3. the material world is the only reality that exists

4. the belief that, whether or not the supernatural exists, we must pretend that it doesn't when studying science

5. *Any two of the following:*

   • Think ahead with an end-goal in mind

   • Use foresight to design a blueprint

   • Collect the proper materials

   • Construct the necessary tools and machines to build

   • Assemble everything in the right order

   • Create new information

6. *Answer should be something like:*

   They examine natural objects for evidence of CSI.

7. *Answer should be something like:*

   ID uses strictly scientific investigations based solely upon the evidence, but creationism starts with religious texts and ends with religious conclusions

8. It is unlikely

9. It matches an independent pattern

10. An intelligent agent

11.

	Complex?	Specified?
Salt Crystal	NO	YES
Hair after waking up	YES	NO
Morse code	YES	YES
Ripples on seashore	NO	YES
Arrangement of garbage	YES	NO
Ink on today's paper	YES	YES

Vocabulary

12. intelligent agent

13. tenet

14. intelligent design

15. materialism

16. Socrates

17. creationism

18. philosophy

19. CSI

Chapter 2

1. Charles Darwin and Alfred Russel Wallace

2. *Origin of Species* / 1859.

3. artificial selection

4. genes (genetics)

5. microevolution

6. macroevolution

7. *Answers may vary.*

8. the grouping of organisms into categories based on features that scientists consider similar

9. species / domain

10. an invalid type of reasoning where the rules or starting assumptions permit only the desired results

11. question assumptions / demand evidence / define your terms / seek the best explanation

12. about 150 years / *Answers may vary*

13. ID does not conflict with microevolution and / or ID does not necessarily conflict with common descent

14. ID contends that many complex biological features could not have been built by random mutations and natural selection

Vocabulary

15. seek more evidence
16. artificial selection
17. Darwinism (original theory)
18. microevolution
19. neo-Darwinism
20. circular reasoning
21. natural selection
22. macroevolution
23. Lamarck
24. evolutionist

Chapter 3

1. 5.88 trillion miles
2. the universe has a cause
3. *Any two of the following:*

   - suggests that a first cause, outside the universe, caused the beginning of the universe
   - forces materialists to confront and explain the evidence for cosmic fine-tuning
   - imposes time constraints on theories of unguided biological evolution, such as Darwinian evolution

4. general relativity
5. giving it a value that supported an eternal universe, and letting philosophical biases influence his science
6. background radiation / COBE satellite
7. 100 billion
8. *Answer should be something like:*

   light waves coming from a receding object are stretched to a lower frequency, and thus shifted down toward the red end of the light spectrum / Hubble discovered a disproportionately high level of red light coming from virtually every galaxy

9.

	age?	size?
Static Model	infinitely old	constant
Steady State Model	infinitely old	expanding
Big Bang Model	finite age	expanding

10. C.

11. *Answers may vary.*

Vocabulary

12. g

13. a

14. d

15. e

16. h

17. c

18. b

19. j

20. i

21. f

Chapter 4

1. finely-tuned (or ideally-suited)
2. the effect in the universe that acts to push galaxies away from one another, accelerating the expansion of the universe
3. gravity
4. *Any three of the following:*

   - strong nuclear force
   - weak nuclear force
   - electromagnetic force

9.

- gravitational constant

- cosmological constant (dark energy)

- initial entropy

5. stars

6. quarter

7. they are dispersed throughout the galaxy OR they are used to manufacture heavier elements inside a new star

8. three

9. about 13.7 billion years

10. initial entropy

11.

a. Universe

b. Galaxy

c. Solar system

d. Sun

e. Jupiter

f. Earth

g. Moon

Vocabulary

12. dark energy

13. strong nuclear force

14. gravitational constant

15. cosmological constant

16. gravity

17. sufficient

18. electromagnetic force

19. necessary

20. initial entropy

Chapter 5

1. multiple universes or multiverse theory

2. chance universe, or theory of chance

3. self-creating universe

4. quantum theory

5. oscillating universe theory

6. *Answer could include:*

- no explanation for the cause of the multiverse

- no reason different universes should have different parameters

- no way to observe other universes

- not falsifiable

- destroys the ability to do scientific investigation

- violates Ockham's Razor

7. *Answer could include:*

- no known mechanism to cause the "bounce"

- loss of gravity caused by the expansion would make it impossible for the universe to contract

- loss of energy at each bounce means limited number of oscillations

8. *Answer could include:*

- no physical explanation does not mean there is no explanation

- theory doesn't address the origin of the quantum vacuum

- quantum theory describes the phenomena, but doesn't explain it

- points to a nonphysical cause

9. *Answer could include:*

- no explanation for cause of fluctuation or the environment that allowed it

- "by chance" is not an explanation

10. *Answer could include:*

- in order for universe to create itself, it would have had to already exist

- laws of physics cannot create the universe, they only describe how it works

Vocabulary

11. For the solution to vocabulary puzzle, see the graphic to the right.

Phrase: Socrates taught follow the argument wherever it leads.

Chapter 6

1. Copernican Principle

2. *Any four of the following:*

- atmosphere composition

- type of star

- stable orbit

- circumstellar habitable zone position

- large moon

- liquid iron core

- liquid water

- plate tectonic activity

- terrestrial planet

3. galactic habitable zone

4. circumstellar habitable zone

5. Jupiter / Saturn

6. 11

7. a planet the size of Mars impacting the Earth

8. moon's size and gravitational pull maintains a stable tilt of Earth's axis, allowing mild seasons

9. Nicolaus Copernicus

10. nitrogen / oxygen

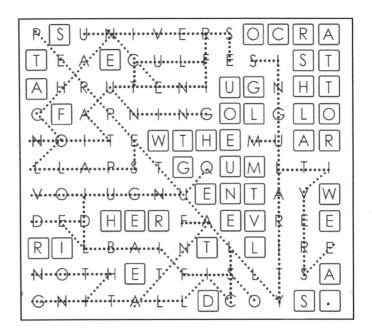

11. *Any two of the following:*

- acts as a "universal solvent," providing an ideal medium for the chemistry of life

- is denser as a liquid than as a solid, meaning ice floats, a property that prevents Earth's oceans from freezing solid

- boils at a high temperature, allowing it to remain liquid over a wide range of temperatures and permitting bodies of water to easily form

- has a high surface tension, fostering many biological processes

- retains heat well, curtailing extreme temperature swings in Earth's climate

Vocabulary

12. b

13. f

14. e

15. h

16. a

17. d

18. g

19. c

20. Solar System Diagram: See Figure 6-3 in the textbook on page 65.

Chapter 7

1. the sudden development of organisms from non-living matter, or life can spontaneously come from non-life

2. proteins

3. the theory that chemicals in nature assembled through blind, unguided, chance chemical reactions to create life

4. 20

5. *Any three of the following:*

- protein production

- transportation of parts within the cell

- energy production

- waste disposal

- replication

- protection from elements outside the cell

6. *Answer could include something like any of the following:*

- used wrong gases

- free oxygen might have destroyed organic molecules

- no geological evidence for soup

- experiment stacked the deck to produce organic molecules

- does not simulate early Earth

- when correct gases used no amino acids are produced

- amino acids are a far cry from life

- does not explain how to link amino acids into proper order to create functional proteins

   / Answers may vary

7. words or sentences

8. ATP synthase

9. Louis Pasteur

10. "warm little pond"

11. yes / free oxygen would cause immediate oxidation, destroying any organic compounds in the "soup"

12. *Any two of the following:*

   - holds cell's contents together to allow for cellular processes to take place

   - protects from harmful molecules and chemical reactions in outside environment

   - allows nutrients and water in

   - allows waste product removal

Vocabulary

13. spontaneous generation

14. primordial soup

15. oxidation

16. Miller-Urey experiment

17. amino acid

18. proteins

19. plasma membrane

20. chemical evolution

21. ribosome

22. ATP

Chapter 8

1. a basic unit of heredity, typically understood as a section of DNA that contains assembly instructions for a particular protein

2. proteins or enzymes / plasma (cell) membrane

3. nucleotide

4. *Answer could include:*

   - RNA can't form without intelligent design

   - RNA can't fulfill the roles of proteins

   - RNA world can't explain the origin of information

   - RNA world can't explain the origin of the genetic code

5. *Answer should be something like:*

   link amino acids in the correct order to produce functional proteins

6. making proteins

7. translation

8. *Answer should be something like:*

   The first step is transcription, where DNA instructions to build a protein are copied into mRNA. The mRNA molecule then travels to the ribosome, where translation occurs. In this step, the ribosome reads, follows, and "translates" the instructions in the mRNA to link a chain of amino acids in the proper order, constructing a protein.

9. *Answers may vary.*

Vocabulary

10. genome

11. tRNA

12. A, G, C, T

13. genetic code

14. transcription

15. RNA world

16. DNA

17. mRNA

18. translation

Chapter 9

1. my theory would absolutely break down

2. natural selection has no effect; it has no reason to preserve the change

3. mutation would tend not to be preserved

4. *Answer should be something like:*

   its special shape allows it to carry iron atoms that attract oxygen

5. a single system that is composed of several interacting parts that contribute to the basic function, and where removal of any one of the parts causes the system to effectively cease functioning

6. *Answer should be something like:*

   irreducibly complex structures cannot evolve in a step-by-step fashion because

they do not function until all their parts are present and functional

7. *Answer should be something like:*

   Proteins must commonly interact or connect through a "hand-in-glove" fit in order to accomplish their cellular functions. Multiple amino acids in each protein must be specifically arranged to give it the proper shape for such a fit. There is too much CSI in many proteins for them to arise by Darwinian mechanisms.

8. 10^{74}

9. *Answer should be something like:*

   in many multicellular organisms, evolving even a modestly complex multi-mutation feature would require greater population sizes and more time than would be available over Earth's history

10. *Answer should be something like:*

   propels bacteria toward food or a hospitable living environment / it could not swim, and might be unable to find food, and might die

11. *Answers may vary.*

Vocabulary

12. b

13. e

14. d

15. g

16. c

17. a

18. f

Chapter 10

1. *Answers may vary.*

ws:

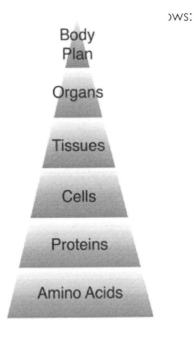

3. Answer could include:

   - limits cellular growth

   - restricts movement

   - protects the plant

   - provides structural stability

4. *Answer should be something like:*

   Chloroplasts capture energy from the sun using chlorophyll. Chloroplasts use this energy to combine water and carbon dioxide to create oxygen and sugars which serve as food for the plant.

5. *Answers should be something like:*

   A process of pre-programmed development where an organism changes its body plan / An organism could not survive complete metamorphosis unless the entire process was fully programmed from the beginning. Such a large jump in complexity requires forethought and planning—things that don't exist in Darwinian evolution.

6. a feature that enables an organism to survive and reproduce in its environment

7. *Animal:* epithelial / connective / muscle / nervous

   Plant: ground / dermal / vascular

8. epithelial

9. dermal

10. nervous

11. xylem

12. *Answer should be something like:*

   it must maintain its high energy level that allows its heart to beat 1000 times per minute, or wings to beat 75 times per second

13. smooth / skeletal / cardiac

Vocabulary

14. g

15. a

16. d

17. e

18. c

19. h

20. i

21. f

22. b

Chapter 11

1. *Any four of the following:*

   - blood capillaries
   - cornea
   - fovea
   - iris
   - lens
   - macula
   - muscles
   - optic nerve
   - pupil
   - retina
   - vitreous gel
   - optic nerve

2. useless

3. The replication of an organism through self-cloning, with no mate required to produce offspring.

4. 100 / 50

5. fallopian tubes

6. *Any five of the following:*

   - ovaries for the production and storage of eggs
   - unique hormones to promote release of eggs (ovulation), and to allow gestation of a baby
   - vaginal entrance for fertilization
   - fallopian tubes
   - placenta to allow implantation of the zygote and nourishment of the developing baby
   - connection to the growing baby through a sturdy umbilical cord
   - uterus to accommodate the growing offspring
   - amniotic fluid
   - hormones to change the female's body in preparation for birth
   - pelvis and birth canal capable of spreading for delivery of the baby
   - mammary glands
   - male hormones
   - testes
   - sperm

7. *Any three of the following:*

   - sexually reproducing organisms face a drop in fitness
   - new beneficial mutations may not be passed on
   - half of the chromosomes must be separated into sex cells
   - sex cells must be protected and preserved until mating

- much energy is spent producing sex organs for mating

- individuals risk not finding a mate

- reproductive systems in males and females must have both physical and biochemical compatibility

- biochemical incompatibilities between parents and offspring can occur

8. villi

9. *Any five of the following:*

   - anus

   - beneficial bacteria

   - bile

   - esophagus

   - gastric juices

   - gall bladder

   - jaw

   - large intestines

   - liver

   - mouth

   - mucus

   - mucous membrane

   - muscles

   - pancreas

   - pancreatic fluid

   - rectum

   - saliva

- salivary glands

- small intestines

- stomach

- teeth

- tongue

- villi

10. alimentary canal

11. mucous membrane

12. *Any three of the following:*

    - auditory

    - circulatory

    - endocrine

    - immune

    - musculoskeletal

    - neurological

    - respiratory

    - urinary

13. *NOTE: will take some outside research on any one of these (or other) systems: auditory, circulatory, endocrine, immune, musculoskeletal, neurological, respiratory, and urinary. Answers may vary.*

14. *Answers may vary.*

Vertebrate Eye Diagram

See Figure 11-1 in the textbook on page 119.

Digestive Tract Diagram

See Figure 11-4 in the textbook on page 124.

Chapter 12

1. often based on the claim that some natural structures are flawed or functionless

2. a structure has features requiring a mind capable of forethought to design the blueprint

3. *Answer should be something like:*

   no, because imperfect structures can still be designed

4. First: we must determine whether the flaw, if real, would actually be an argument against intelligent design / Second: we must investigate whether the flaw is real

5. they act like fiber optic cables to channel light through the optic nerve wires directly to the photoreceptor cells

6. *Answers may vary.*

7. *Answers may vary.*

8. biomimetics

9. *Answer should be something like:*

   Yes and no. Darwinian evolution is the best explanation for blind eyes on cave salamanders. But ID allows that Darwinian evolution can cause loss-of-function. ID is more interested in how complex features like functional eyes are gained in the first place than how they can be lost.

10. Anything between 80% and 100% / *Any three of the following:*

    • repairing DNA

    • assisting in DNA replication

    • regulating gene expression

    • aiding in folding and maintenance of chromosomes

    • controlling RNA editing and splicing

    • helping to fight disease

    • regulating embryological development

11. *Answer should be something like:*

    No, because the appendix serves as a storehouse for beneficial bacteria in our intestines that aid in the breakdown of food, and serves a variety of immune-related functions.

12. *Answer should be something like:*

    Because function and purpose is continually being discovered for the so-called vestigial organs.

13. *Answers may vary.*

Vocabulary

14. panda's thumb

15. optic nerve

16. appendix

17. junk DNA

18. SLN

19. dysteleology

20. vestigial

21. probiotics

22. pseudogenes

23. equivocation

Chapter 13

1. universal common ancestry

2. similarity between organisms is the result of inheritance from a common ancestor

3. morphology

4. two or more species independently acquiring the same trait, supposedly through Darwinian evolution

5. *Answers may vary.*

6. comparing similarities in DNA, RNA or protein sequences in different organisms

7. comparing physical characteristics, such as anatomical and structural similarities of different organisms

8. a process by which microorganisms can obtain genes by sharing and swapping genes with their neighbors

9. Any two of the following:

   • there is no evidence that HGT occurs with all the essential genes

   • it isn't observed among higher organisms like plants and animals

   • uses circular reasoning to claim that conflicts between phylogenetic trees are evidence for HGT

   • ignores the possibility that conflicts in the tree of life exist because common descent is false

10. similarity of structure and position, but not necessarily function

11. taking evidence for common design, and mistakenly calling it evidence for common descent

12. *Answer should be something like:*

    Common design is an entirely reasonable explanation for functional anatomical similarities. Why must a designer always start over when a useful design already exists?

Vocabulary

13. c

14. e

15. a

16. b

17. d

18. g

19. f

Chapter 14

1. the development of an organism (ontogeny) replays (recapitulates) its evolutionary history (phylogeny) / vertebrate embryos

do not replay their supposed earlier evolutionary stages

2. Ernst Haeckel

3. vertebrate embryos / *Answer should be something like:* Haeckel's drawings obscured the differences between the earliest stages of embryonic development, making the embryos look more similar than they actually are

4. developmental hourglass

5. *Any three of the following:*

   • body size

   • body plan

   • growth patterns

   • timing of development

6. the supposed rapid diversification of a species after entering an empty habitat, or niche.

7. *Answers may vary.*

8. *Any two of the following:*

   • at most it would demonstrate microevolution

   • there is controversy over whether peppered moths normally rest on tree trunks in the wild

   • many textbook photographs showing peppered moths on tree trunks are staged

   • moths are now reverting back to a lighter color, so at best this is an example of oscillating selection, where there is no net evolutionary change over time

Vocabulary

9. e

10. d

11. c

12. i

13. g

14. h

15. f

16. b

17. a

18. j

Chapter 15

1. *Any four of the following:*

   • Cambrian explosion

   • fish

   • land plants

   • "big bloom" of angiosperms

   • mammals

   • birds

   • genus *Homo*

2. *Answers may vary / Answers may vary.*

3. the extreme imperfection of the fossil record

4. the artifact hypothesis

5. *Answers may vary.*

6. the idea that most evolution takes place in small populations over relatively short geological time periods, followed by periods of stasis

7. *Answer should be something like:*

    It deprives Darwinism of two important advantages: large populations and long timescales, allowing too few roles of the dice for beneficial mutations to arise

8. *Answer should be something like:*

    The theory that changes to the master genes that control the development of an organism, such as *Hox* or *homeobox* genes, can cause large, abrupt changes in body plans

9. *Answers may vary but can be something like one of the following:*

    • changes in developmental genes affect many others, making most mutations lethal

    • *Hox* genes do not encode proteins that build body parts—they merely direct the genes that encode body parts, and cannot create truly novel structures

    • the best examples of change produced by evo-devo mechanisms are meager and often entail loss, rather than gain of function

10. *Answers may vary / Answers may vary.*

Vocabulary

11. punc eq

12. evo-devo

13. big bloom

14. Cambrian explosion

15. artifact hypothesis

16. *Hox* genes

17. explosion

18. fossils

Chapter 16

1. a type of carnivorous, bipedal dinosaur with small forelimbs

2. birds

3. it had feathers like birds, but also features generally associated with reptiles—claws on its forelimbs, teeth, and a bony tail

4. *Any four of the following:*

    • cold-blooded (ectothermic) to warm-blooded (endothermic)

    • slow metabolism to fast metabolism

    • three-chambered heart to four-chambered heart

    • no feathers to feathers

    • typically abandon young to instincts to care for young

    • breathe using diaphragm to breathe using air sacs

Fossil	Millions of years ago
Tetrapod tracks (reported in 2010)	397
Hypothetical fish ancestors of tetrapods	390
Tiktaalik roseae	375
Tetrapod body fossils	360

5. *Answers may vary.*

6. feathered / dinosaurs / fossils

7. Chart should be completed as follows:

8. tetrapod tracks were discovered in Poland that predated *Tiktaalik* by about 20 million years; *Tiktaalik*'s presumed tetrapod descendants now appear 20 million years before *Tiktaalik* itself appears

9. Any two of the following:

   • it's an imaginary lineage

   • in the actual fossil record, some earlier horses are larger than later ones

   • fossils span different continents, separated by vast expanses of ocean

   • both location and sequence conflict with the tidy evolutionary series often portrayed

   • the horse body plan does not significantly evolve over the lineage

10. *Any one of the following:*

    • Emergence of a blowhole

    • Modification of the eye for underwater vision

    • Ability to drink sea water

    • Forelimbs transformed into flippers

    • Modification of skeletal structure

    • Ability to nurse young underwater

    • Origin of tail flukes and musculature

    • Blubber for temperature insulation

    / Answers may vary.

11. *Answer should be something like:*

    On the rare occasions when there actually are fossils that show potentially intermediate traits, unguided neo-Darwinian evolution is invalidated by the short amount of time allowed by the fossil record.

Vocabulary

12. d

13. b

14. e

15. g

16. a

17. c

18. h

19. f

Chapter 17

1. the study of human origins

2. *Answers may vary but could include:*

   • because conclusions of emotional significance to many must be drawn from extremely paltry evidence

   • because much of paleoanthropology is not based on objective criteria, but on subjective interpretations of scant fossil evidence

3. humans and chimpanzees, and any extinct

organisms leading back to their presumed most recent common ancestor

4. a

5. c and / or e

6. a

7. f

8. b (c, d, and e could also be correct)

9. a

10. c

11. e

12. f

13. d

14. *Answers may vary.*

15. *Answers may vary but may include something like any two of the following:*

- the 1% difference reflects only base substitutions, not the many stretches of DNA that have been inserted or deleted in the genomes

- some aspects of our genome are very different from chimps, such as the y-chromosome

- genetic comparisons typically focus only on differences in protein-coding DNA and ignore differences in noncoding DNA

- actual differences amount to at least 35 million base-pair changes, 5 million insertions and deletions in each species, and 689 extra genes in humans

- depending on how you do the analysis, human and chimp genome similarity might be as low as 70%

- even if the claim was true, it would not demonstrate human/chimp common ancestry; functional genetic similarities could be explained by common design

16. *Any three of the following:*

- extreme altruism and acts of human kindness

- saving strangers trapped inside a burning vehicle

- voluntary poverty

- celibacy

- martyrdom

- charitable abilities

- artistic abilities

- intellectual abilities

- composing symphonies

- discovering quantum mechanics

- building cathedrals

   Other similar human activities could also be correct.

17. *Answers may vary but could include something like:*

   the ability is far beyond the basic requirements of natural selection to survive and reproduce

Vocabulary

18. human exceptionalism

19. *Australopithecus*

20. kin selection

21. reciprocal altruism

22. Neanderthals

23. paleoanthropologist

24. *habilis*

25. *Homo*

26. Hominidae

Chapter 18

Answers to all questions may vary.

Chapter 19

1. *Answers should be something like:*

   Straw Man: Misrepresenting an opponent's argument, and then attacking the false version.

   Professional Intimidation: Citing one's educational training or professional accomplishments to silence another's viewpoint.

   Genetic Fallacy: Attacking the origin of an argument rather than the argument itself.

   Argument from Popularity: Urging people to believe an idea because it is popular.

   False Choices: Portraying two options as mutually exclusive when they really are not.

2. ridiculed / accepted as being self-evident

3. *Inherit the Wind*

4. John Scopes

5. *Expelled: No Intelligence Allowed*

6. ID Movement / Darwin Lobby

7. 78%

8. the tendency to attribute anything not understood to God or the gods

9. *Answers may vary.*

Vocabulary

10. j

11. c

12. a

13. b

14. h

15. d

16. k

17. i

18. e

19. f

20. g

Chapter 20

1. The four steps are:
   - observation
   - hypothesis

- experimentation

- conclusion

2. *Answers may vary.*

3. *Any two of the following:*

   - make falsifiable predictions

   - undergo peer-review

   - use only materialistic explanations

   / Answers may vary.

4. a process by which scientists double-check the work of their peers to make sure it was done correctly

5. *Any two of the following:*

   - Natural structures will be found that contain many parts arranged in intricate patterns (including irreducible complexity) that perform a specific function—high CSI.

   - Forms containing large amounts of novel information will appear in the fossil record suddenly and without similar precursors.

   - Genes and other functional parts will be commonly re-used in different organisms.

   - Much so-called "junk DNA" will turn out to perform valuable functions.

   / Answers may vary.

6. *Answers may vary.*

7. *Kitzmiller v. Dover / Answers should be something like any two of the following:*

   - adopted a false definition of ID by claiming that ID requires supernatural causation, that it is creationism, and is a negative argument against evolution

   - denied the existence of pro-ID, peer-reviewed, scientific publications and research

   - adopted an unfair double-standard of legal analysis where religious implications, beliefs, and motives count against ID but never against materialist theories

   - tried to define science, settle controversial scientific questions, and explain the proper relationship between evolution and religion

   - attempted to turn science into a voting contest by claiming that popularity is required for an idea to be scientific

   - tried to refute ID, but a court ruling cannot negate the scientific evidence pointing to intelligent design

8. to restrict free speech and intellectual inquiry, stopping people from discussing non-evolutionary views

9. to expand intellectual inquiry and free speech rights, defending academic freedom to debate evolution

10. the evidence / *Answers may vary.*

11. *Any two of the following:*

    - think critically and question assumptions

    - continue to learn

    - get involved

12. a

13. d

14. b

15. c

16. f

17. e

18. zero-concession policy / absolutely
    nothing / clarify our arguments / appeal to
    the undecided middle

Vocabulary

19. a

20. c

21. b

22. h

23. f

24. d

25. g

26. e